Oxford Applied Mathematics
and Computing Science Series

Oxford Applied Mathematics
and Computing Science Series

I. Anderson: *A First Course in Combinatorial Mathematics* (*Second Edition*)
D. W. Jordan and P. Smith: *Nonlinear Ordinary Differential Equations* (*Second Edition*)
D. S. Jones: *Elementary Information Theory*
B. Carré: *Graphs and Networks*
A. J. Davies: *The Finite Element Method*
W. E. Williams: *Partial Differential Equations*
R. G. Garside: *The Architecture of Digital Computers*
J. C. Newby: *Mathematics for the Biological Sciences*
G. D. Smith: *Numerical Solution of Partial Differential Equations (Third Edition)*
J. R. Ullmann: *A Pascal Database Book*
S. Barnett and R. G Cameron: *Introduction to Mathematical Control Theory* (*Second Edition*)
A. B. Tayler: *Mathematical Models in Applied Mechanics*
R. Hill: *A First Course in Coding Theory*
P. Baxandall and H. Liebeck: *Vector Calculus*
P. Thomas, H. Robinson, and J. Emms: *Abstract Data Types: Their Specification, Representation, and Use*
D. R. Bland: *Wave Theory and Applications*
R. P. Whittington: *Database Systems Engineering*
J. J. Modi: *Parallel Algorithms and Matrix Computation*
P. Gibbins: *Logic with Prolog*
W. L. Wood: *Practical Time-stepping Schemes*
D. J. Acheson: *Elementary Fluid Dynamics*
L. M. Hocking: *Optimal Control: An Introduction to the Theory with Applications*
S. Barnett: *Matrices: Methods and Applications*
P. Grindrod: *Patterns and Waves: The Theory and Applications of Reaction–Diffusion Equations*
O. Pretzel: *Error Correcting Codes and Finite Fields*
D. C. Ince: *An Introduction to Discrete Mathematics, Formal System Specification, and Z* (*Second Edition*)

D. [illegible]

Open [illegible] ...agement

An Introduction to Discrete Mathematics, Formal System Specification, and Z

Second edition

CLARENDON PRESS · OXFORD

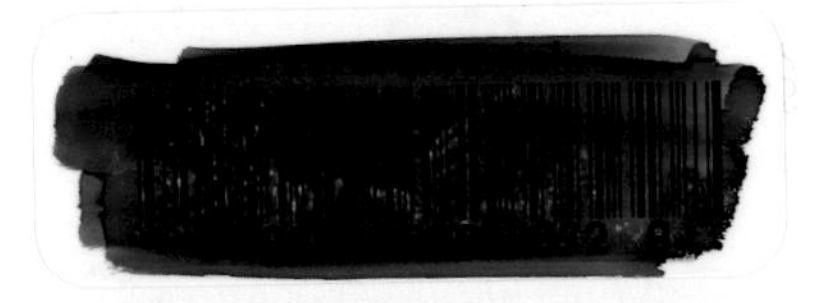

Oxford University Press, Walton Street, Oxford OX2 6DP
Oxford New York Toronto
Delhi Bombay Calcutta Madras Karachi
Kuala Lumpur Singapore Hong Kong Tokyo
Nairobi Dar es Salaam Cape Town
Melbourne Auckland Madrid
and associated companies in
Berlin Ibadan

Oxford is a trade mark of Oxford University Press

Published in the United States
by Oxford University Press Inc., New York

First published as An Introduction to Discrete Mathematics and
Formal System Specification, 1988
Reprinted 1993

A catalogue record for this book is available from the British Library

Library of Congress Cataloging in Publication Data
Ince, D. (Darrel)
An introduction to discrete mathematics formal system specification, and Z / D. C. Ince. – 2nd ed.
(Oxford applied mathematics and computing science series)
Updated ed. of: An introduction to discrete mathematics and formal system specification. 1988.
Includes bibliographical references.
1. Computer science–Mathematics. 2. System design. I. Ince, D. (Darrel). Introduction to discrete mathematics and formal system specification. II. Title. III. Series.
QA76.9.M35153 1992
005.1'1–dc20
ISBN 0–19–853837–5 (hbk.)
ISBN 0–19–853836–7 (pbk.)

Printed in Great Britain by
Bookcraft (Bath) Ltd,
Midsomer Norton, Avon.

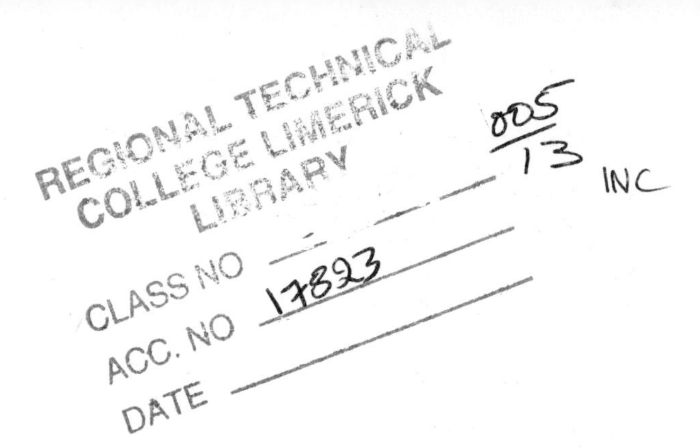

Preface

This is the second edition of a book which was published when formal methods of software development were still in their comparative infancy. The aim of this edition is the same as that of the previous edition: to provide an introduction to discrete mathematics and formal system specification that can be accessed by anyone with a knowledge of mathematics attained by study at school to the age of sixteen. There are two audiences which I hope this book will attract. The first audience is staff working on software projects who feel they require a knowledge of formal methods—particularly those staff in the defence sector, where the recent defence standard 00-55 has made formal methods virtually mandatory. The second audience is students studying for computer science and software engineering degrees or diplomas who require a knowledge of discrete mathematics before progressing to advanced courses.

This book differs from its predecessor in a number of ways:

- The material on inference has been moved to an earlier point in the book. In the previous edition it was placed in the chapter on predicate calculus. I have revised my opinion of where to put it, based on my and other academics experiences of teaching the topic. It is now placed in the chapter on propositional calculus.

- I have added two new chapters. First, I have added a short chapter containing a number of small examples in Z. This forms a bridge between the early descriptions of Z in the book and the large case study that appears in Chapter 12. Second, I have included a chapter on design and Z. When the first edition was written I felt that our knowledge of formal design was such that a chapter on the topic would be premature. Happily this has changed. This chapter also gives a solid example of how reasoning is useful in formal software development—something which was absent in the first edition.

- The Z notation used in the book now conforms to the *de facto* standard.

- There are eight new worked examples.

- The book has been produced using LaTeX, rather than by conventional typesetting means. This, I believe, has improved its overall appearance. Although the struggle to overpower the page breaking algorithms in LaTeX almost killed me.

In the same way that it is important to provide an upgrade path for software, I have designed this edition in such a way that anyone who has taught classes using the first edition can easily use this edition. All that is required is to teach the first 12 chapters. Chapter 13 can either be used as additional reading, or can be formally integrated into the version of the course based on the first edition, when the time comes for a rewriting of the course.

Finally I want to thank Michael Lutz of the Rochester Institute of Technology, Jonathon Jacky of the University of Washington, Paul Samet of University College London, Mark Priestley of the Polytechnic of Central London and Ursula Martin of Royal Holloway College, for the advice and encouragement that they have given me during the writing of this edition. Needless to say any errors, ambiguities and infelicities that remain are solely my fault.

Milton Keynes D.I.
May 1992

Contents

1

Commercial software development

Aims

- To describe how modern software projects are managed and organized.
- To describe some problems currently encountered in software development.
- To describe some misconceptions about the use of mathematics on a software project.

1.1 The software life cycle

Many readers of this book will be familiar with programming only as a software development activity. This chapter provides an introduction to a number of the more important software development activities and places the rest of the book firmly in the context of an engineering approach to software.

Producing a major software system with a large number of staff, each of whom will have a wide range of competence and skills, is one of the most complex activities the human race currently undertakes.

About twenty years ago programs were very small. Perhaps the largest occupied a few hundred lines of code. Software systems now contain hundreds of thousands of lines of code and a much more disciplined engineering approach to software development has to be adopted. It involves splitting the development process into a series of phases, each phase being associated with a distinct activity. These phases are: **requirements analysis**, **system specification**, **system design**, **detailed design**, **coding**, **integration**, **operation**, and **maintenance**. Figure 1.1 shows a summary of the activities of the software life cycle.

1.1.1 Software requirements analysis and system specification

The software life cycle starts with requirements analysis and system specification. The input to the requirements analysis and specification is a **customer statement of requirements** and the output is a **software specification**. The former is the customer's idea of what a system is to

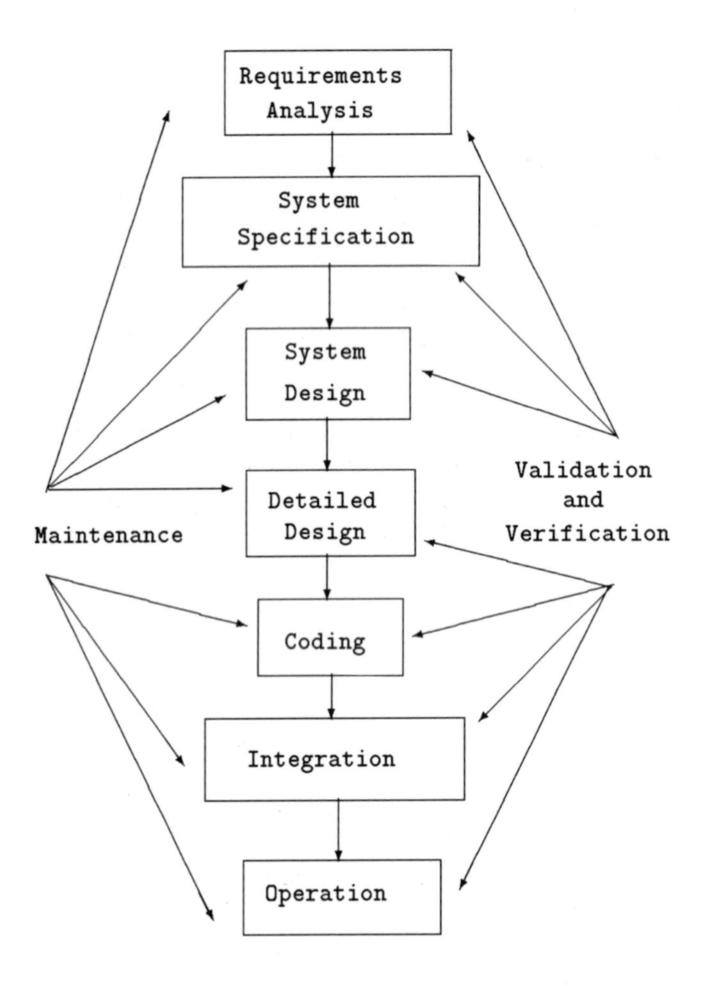

Figure 1.1 Activities of the software life cycle

do; the latter is an unambiguous version of the former. A sample from a customer statement of requirements is shown below.

7 Storage of water level readings

7.1 The software should store the hour-by-hour readings from sixteen sensors for the last three months.

7.2 Data will be held on magnetic tape which should be continually mounted.

7.3 A reading of zero indicates that a malfunction in a reservoir sensor has occurred. If more than six malfunctions occur in a day, then an error should be displayed.

8 Query facilities

8.1 The software should enable the average daily water level for the past three months to be displayed for any day.

8.2 The software should enable the average monthly water level for the past three months to be displayed for any of these months.

8.3 The response to these queries should be of an acceptable duration.

The process of requirements analysis and system specification consists of removing errors in the statement of requirements. The end-product of this activity will be a system specification. Such a document will contain both **functional** and **non-functional** requirements.

A functional requirement is a statement of what a software system is to do. Typical functional requirements are shown below.

When the DISPLAY command is executed the program should calculate and display the average of the last 10 transducer readings and display this figure on the operator's console.

The monitoring subsystem should produce at the end of every minute an up-to-date display of reactor pressures for all those reactors connected to the primary circuit.

Non-functional requirements are those which are concerned with practical constraints upon the software developer. Typical non-functional requirements are shown below.

The command and control subsystem should always be resident in main memory.

The response to MONITOR commands should be less than 0.4s—even at peak processing time.

The system should be programmed in COBOL74.

A critical determinant of project success is the system specification. Poor specifications have led to overruns and budget problems. Early attempts have led to graphical notations [Weinberg, 1980, Schoman and Ross, 1977], special-purpose languages [Bell *et al.*, 1977, Teichrow and Hershey, 1977], and notations based on discrete mathematics [Bjorner and Jones, 1982, Guttag and Horning, 1978].

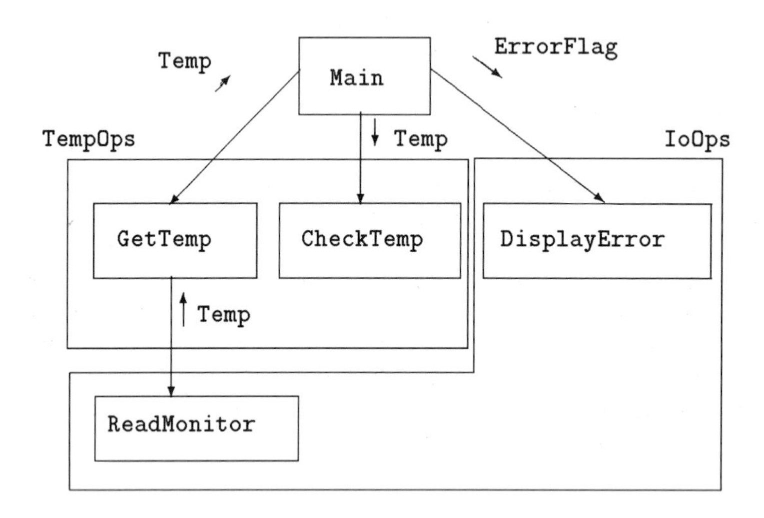

Figure 1.2 A structure chart

1.1.2 System design

System design is the process of defining an architecture which satisfies the system specification. It involves the designer partitioning the software into functionally related groupings known as **modules**. Such modules will consist of constants, variables, types and program units (functions and procedures) which provide resources to carry out a series of related tasks.

The result of the system design phase is a system architecture which expresses the relationships between individual program units in a proposed system. A very simple example is shown in Figure 1.2. It is expressed in a graphical notation known as a structure chart and shows the design of a temperature monitoring system. This reads temperatures from a thermocouple and displays an error message when an out-of-range value occurs. The system design uses four program units. Two of these, *GetTemp* and *CheckTemp*, are part of the module *TempOps*, while the program unit *DisplayError* is part of the module *IoOps*. *Main* represents the main program which calls the program units. The interface between the program units is represented by the data items *Temp* and *ErrorFlag*. As with all system designs the one shown in Figure 1.2 is hierarchic: the program units *GetTemp*, *CheckTemp*, and *DisplayError* are subordinate to *Main* by virtue of the fact that they are called by *Main*.

A second task in system design is to design the data base which the developed software will use. This involves the specification of: the structure of the data base; the access methods used; and the control, security, and protection features that are to be employed.

1.1.3 Detailed design

Detailed design is the process of transforming a system design into a form in which it can be given to a programmer and implemented. It involves the design of the individual algorithms that are to be used in the developed system.

An example of a detailed design notation is shown below. It represents the design of a program unit *addup* in a program design language [Cain and Gordon, 1975]. Such a language consists of the normal control constructs found in a programming language augmented with natural language. The function of *addup* is to read a series of 60 temperatures and find the sum of those between zero and one hundred degrees.

```
procedure addup(sum)
Initialize sum
for i to 60 do
   Read a temperature
   if the temperature is between 0 and 100 then
      add the temperature to sum
   endif
endfor
endprocedure
```

As well as refining the system design, the data-base design should also be refined. This involves expanding entities such as *Salesman*, *HostilePlane*, *BudgetDeficit*, and *ReactorRecord* which were defined in abstract terms during system design into more concrete descriptions. For example, a detailed design which dealt with an entity called *City* might define it as an alphabetic field of a record which would contain no more than 25 characters.

1.1.4 Coding

Coding is also known as programming. It consists of processing a detailed design and producing program code which reflects that design. This is an almost automatic process, since the detailed design phase will have defined the control structures and processing that is to occur in the final system.

1.1.5 Integration

Integration is the name given to the process of taking a coded program unit or module and adding it to a software system during the course of construction. A developer will integrate modules and program units into a software system gradually. Each time that a module or program unit is integrated the developer will have a high degree of confidence that errors present will occur only in the module or unit that has been integrated.

1.1.6 Maintenance

Maintenance is the term given to the changes made to a software system after it has been put into operation. There are a number of reasons for these changes. First, errors will have been committed which have only been discovered during operation. Second, the environment in which a software system is placed often changes during operation. For example, a new version of an operating system may be installed or a new instrument interface may be added to an existing system. Third, the requirements for the implemented system may change. For example, a customer who currently uses an accounting package may require changes to that package in order to cope with new rules for reporting tax payments.

The increasing pressures of software maintenance place a requirement on the software engineer to devise better notations for system specification: first, in order to reduce maintenance activities due to residual errors which arise from poor specification; second, in order to help the developer understand an existing system before modifying it in response to maintenance changes which are not due to errors.

1.1.7 Validation and verification

At certain times during a software project the evolving software product will be described by documents which specify a set of properties of the software: the customer statement of requirements will describe the customer's view of what the software should do; the system specification will describe the software developer's view of what the software will do; the system design will describe the gross architecture of a software system which will satisfy the system specification; the detailed design will represent the final design of a software system; and, finally, the program code will represent the realization of the system. Each of these documents will be expressed in separate notations. A major activity that occurs throughout a software project is the detection and rectification of errors that occur in these documents. This is known as **validation and verification**.

Verification is concerned with checking that one phase in the software project accurately reflects the intentions of the previous phase. Thus, checking that the detailed design of a program unit matches its specification is an example of verification. **Validation** is concerned with checking that the product of a phase matches customer requirements. Thus, system testing, the process of executing a complete system to check that it meets user requirements, is an example of validation.

The process of validation can be characterized as answering the question 'are we building the right product?' and verification as those activities dedicated to answering the question 'are we building the product right?' [Boehm, 1981]

Testing is a subset of validation and verification. It consists of executing the program code of a software system in order to carry out validation or verification. Since testing requires the use of program code, it can only occur late in the software project. Normally the term 'testing' is associated with the multiple execution of a software system with test data. However, new techniques such as symbolic execution [Clarke, 1976], static analysis [Osterweil, 1983], and dynamic analysis [Miller, 1977] have enlarged the range of testing strategies. Nevertheless, the majority of software projects still rely on exercising a software system with data.

1.2 Current problems in software development

A number of major problems afflict software projects. Many arise from the unsatisfactory nature of the notations used to describe a software product as it progresses through its life cycle. Typically, a developer processes a statement of requirements in natural language, constructs a system design in a graphical notation heavily annotated with natural language, refines this design into a detailed design consisting of a set of control constructs and natural language, and then codes the system in a programming language.

In the next chapter the deficiencies of natural language will be studied in detail. At this stage it is sufficient to say that it is a medium in which it is very difficult to be precise, to state essential properties of a software system without being cluttered by detail, and to reason with. Unfortunately, those notations which are most heavily dependent on natural language are those which occur at the beginning of the software life cycle where undetected errors can lead to massive overruns or even total cancellation.

A second problem occurs because of the fact that developers use a variety of notations in a software project. Often a different notation is used for each phase. This results in errors at the interface between phases.

In order to overcome many of the problems associated with current software development an increasing number of computer scientists have been turning to mathematics as a notation for describing a software product. The use of mathematics to aid in software development is known as 'formal methods'.

1.3 Formal methods of software development

The term 'formal methods' is a general description of the use of mathematical notations such as logic and set theory to describe system specifications and software designs together with techniques of validation and verification based on mathematics.

Formal methods have their root in research on program proving during the late 1960s and early 1970s. Researchers attempted to define the function of a software system using mathematics and then prove that the program code of the system met the specification.

The ultimate aim of formal methods researchers in the 1970s was the development of fully automatic proof systems. Such systems would be able to decide automatically whether a program met its specification without any testing taking place. Automatic proof systems were felt necessary because manual proof tended to be error-prone and complex. In many cases the actual proof exceeded the size of the program. Unfortunately, the complexity of the proof procedure defeated the developers of automatic proof systems, although a number of semi-automatic systems currently exist [Gerhart *et al.*, 1980, Luckham *et al.*, 1979, Constable and O'Donell, 1978].

The comparative failure of program proving has not diminished interest in formal methods. What it has done is to move the emphasis away from formal proof towards the notations used, and towards informal demonstrations of correctness. Research in the past 10 years has concentrated on mathematics-based languages for system specification and design, together with the development of methods for the transformation of a software system expressed in such notations into a concrete software system.

Before examining the inadequacies in detail it would be useful to examine what formal notations can and cannot do for the software developer and also state some of the many misconceptions that surround the use of formal methods on software projects.

Formal methods eliminate the need for testing

Wrong, there will always be a need to test a software system. First, the customer will always require a demonstration with live data to show that a system works before he or she takes responsibility for it. Therefore, the developer will always have to carry out system testing and the developer and customer jointly will have to carry out acceptance testing.

There will always be a need to perform some testing during development in order to carry out a little empirical checking that errors have not been made in the mathematics used to describe the system. What can be said is that formal methods greatly reduce the amount of conventional testing required and enables errors to be discovered much earlier in the software life cycle than the coding phase.

Formal methods eliminate the need for natural language

This is another misconception. Paradoxically, it is often stated by both the proponents and opponents of formal methods in order to support their case. The software developer will always start with a natural language statement of requirements couched in terms of the application area. To expect otherwise would be grossly unrealistic. Also, formal specification documents will contain natural language. However, their role will be different from the role of natural language in current-project documents. In formal specification documents natural language is the medium used to reason, validate, and to verify. What can be said is that natural language in formal methods plays a subsidiary role to the mathematical notation rather than, at present, playing the only role.

You need a Ph.D. in mathematics to understand formal methods

Many people believe that the use of mathematics in software development is just the invention of academics with little else to do but produce techniques and methods which represent an unattainable ideal. They believe that to employ mathematics on a software project ignores the fact that many staff will be unable to use and understand such a notation. There are a number of answers to this. First, academics are in the business of putting forward ideals and criticizing the *status quo*. That is part of their job description. However, the ideals of today often end up being the working practices of tomorrow. Computer science and mathematics are full of examples of this. Research into formal language theory is now being used in the development of fast compilers for programming languages. Academics have produced results in number theory which are used in highly secure cryptographic systems. In software development academics were in the forefront of the structured programming movement.

A more serious objection is that the mathematics used is too difficult. At first sight this seems true. A formal specification document will contain Greek letters, inverted capital letters and symbols such as $\mathbb{P}$, $\cup$, and $\exists$ which are unfamiliar to the majority of computer staff. However, a formal notation is just another notation in the sense that a programming language is just a notation. I can still remember the sense of apprehension that I felt when I saw my first formal specification. It was exactly like the sense of apprehension that I felt when I saw my first program in a language whose use I now regard as second nature.

What can be said is that the notations used in formal specifications are not opaque but just unfamiliar; with sympathetic teaching such notations can be used by the majority of software staff. It has been the weakness of formal methods workers that they have not generated enough tutorial level material for readers without a mathematical background. It is the aim of this book to do just this. It teaches the mathematics in a relaxed way and introduces the application of such mathematics in one specification language.

If I adopt formal methods, then I will have to change my development methods

Again this is a misconception. Formal methods of software development can be slotted into any life cycle oriented method of software development. Of course, it will mean a developer abandoning current notations, not a task to be underrated. However, many techniques of validation and verification become enhanced or are more feasible with a formal notation. For example, reviews are much more effective when the participants can argue about correctness more effectively. The use of formal notations also enables prototyping to be carried out more easily since a formal specification is an exact document expressed in a language with precise semantics. Such a description can be made executable or can be converted into an executable model.

1.4 Further reading

Good descriptions of why mathematics is a good medium for software documentation can be found in [Meyer, 1985] and [Ince, 1988]. A very level-headed discussion of the capabilities and limitations of formal methods can be found in [Craigen and Summerskill, 1990]. A good introduction to a variety of formal methods of software development can be found in [Cohen *et al.*, 1986]. The authors of this work made the potentially catastrophic decision to include some deep mathematics early in the book, however, if you ignore this and persevere with the remainder of the book you will discover an excellent short introduction to the subject. Finally, [Humphrey, 1989] contains an excellent description of how modern software projects should be organized and managed. A major error made by many of the proponents of formal methods is that all you need to solve the software problems that we face is a stiff dose of mathematics. It also needs good project management and quality assurance, and this book is the best description of these topics that I know of.

2

Customer requirements and specification

Aims

- To examine the problems that occur with natural language based documents.
- To outline what makes a good specification.
- To detail a simple procedure for requirements analysis and system specification.
- To show how mathematics is a good medium for specification.

2.1 Introduction

The aim of this chapter is to examine some of the faults found in customer statements of requirements and software specifications written in natural language. The purpose in doing this is twofold. First, the document given to staff who produce the software specification is the customer statement of requirements. It is important that such staff realize the nature of such documents. Second, the inadequacies of natural language as a specification medium are most manifest in statements of requirements and software specifications. Before attempting to describe the use of mathematics in system specification, it will be useful to describe the faults that it should eliminate.

The starting point in the vast majority of software projects is the customer statement of requirements. This is an informal description of the properties of a new system or an addition to an existing system. Normally it is written in natural language; it is expressed in terms of the application rather than computing terms. Thus, it will contain words such as: 'invoice', 'reactor', 'valve', and 'radar range' rather than words such as: 'parameter', 'module', and 'real type'.

The statement of requirements will normally require considerable processing before it becomes a viable document for development staff. The document produced is the system specification. It should be clear, consistent and unambiguous as it forms the major reference document for subsequent activities in the software project. For example, it will be extensively used to generate acceptance tests which are used by the customer and the developer to check that a completed system meets user requirements.

2.2 The content of the statement of requirements

The statement of requirements will contain statements which detail what a system is to do. They are known as **functional requirements**. Some simple examples of functional requirements follow.

> When the SUMMARY command is typed the average reactor temperatures of all current functioning reactors is displayed.
>
> The purpose of the command and control system is to monitor the activities of mobile units which are organized according to NATO directive NS/TC/107/ALL.
>
> The function of the report subsystem is to produce management reports on salesman performance over the current financial year.

As well as statements of function the statement of requirements will contain directives which will constrain the developer in subsequent activities in the life cycle. These are often referred to as **non-functional requirements**. Typical examples of such requirements follow.

> The response to the CLOSEDOWN command should be no slower than four seconds.
>
> The system should be capable of being restored to normal within one minute of a local hardware fault in any duplicated hardware unit.
>
> The programming language used should be ISO standard Pascal.
>
> All access to the 3340 filestore should be via operating system calls.

All the above statements constrain subsequent activities. The first requirement directs the system designer to develop an architecture which is efficient enough to respond to a command in 4 seconds. The second non-functional requirement directs the system designer to include a facility for dumping the state of the system to a filestore. The third and fourth non-functional requirements constrain the staff involved in implementation.

A statement of requirements will also contain sentences which are **design directives** or **implementation directives**. Typical examples are

> The monitoring system should consist of a validation subsystem, a command subsystem, and a reactor monitoring subsystem.
>
> The function of the DISPLAY command is to retrieve the salesman names contained in the main indexed sequential file and display them on the VDU (visual display unit) from which the command was typed.
>
> When the operator types the TRAFFIC command all the traffic statistics for the last three hours will be transferred to the host computer's main memory and subsequently displayed on the VDU on which the command was typed.

The first statement is an outline of a system design. The second statement constrains the designer to use one particular file-based data structure. The third statement constrains the designer to use an algorithm which transfers *all* processed data into main memory rather than an algorithm which carries out the transfer in discrete blocks.

Such statements should be discouraged. They may overconstrain the system designer and result in a non-optimal system which, although it carries out its intended function, is not the best solution. For example, the third design directive above may result in a system which just satisfies all its requirements. If this design directive were eliminated and data transferred in chunks, then main memory could be released and used to hold more memory-resident software and hence increase response time.

Statements of requirements are not usually written by staff who are good designers. Such staff are unable to judge the impact of such instructions on the overall design of the system. However, there are occasionally good reasons for statements of requirements containing design or implementation directives. For example, the second design directive could have been written because the salesman file may have already been implemented as part of an existing system.

A statement of requirements will often contain **goals**. These are statements which guide the software developer where choice exists. Typical goals are

> In developing the system the response time should be minimized.
>
> If possible, the amount of main memory used by the system should be as small as possible.

Such statements are used by the system designer in selecting alternative designs and making trade-offs between properties such as response time and memory utilization.

Finally, a statement of requirements will contain references to data used by a proposed system. Normally these references are manifested in statements such as

> An employee record consists of a name, address, tax code, department, and current salary.
>
> The system should maintain the current status, temperature, and pressure of all functioning reactors.

or indirectly as

> The command PLANE-STATUS, when executed, will display at the master console the identity of each squadron together with the combat status and the number of planes undergoing routine maintenance.
>
> On receipt of the ready signal the medium-speed communication line should respond with the identity of the user currently logged into the communication terminal attached to the line.

2.3 Deficiencies in specifications

Life would be uncomplicated for the software developer if both the statement of requirements and the system specification consisted of a series of sections marked

- Functional requirements;
- Non-functional requirements;
- Goals;
- Data requirements;
- Implementation and design directives

each of which were consistent, unambiguous, and complete and where statements of requirements would be expressed in user terms and the system specification in computing terms. Unfortunately, this very rarely happens. The purpose of this section is to outline how reality deviates from the ideal.

In general a statement of requirements will be vague, contradictory, incomplete, and will contain functional requirements, constraints, and goals randomly mixed at different levels of abstraction. Often it will either have a very naive and over-ambitious view of the capabilities of a software system or a view which was current a few decades ago.

Vagueness

A statement of requirements can be a very bulky document and to achieve a high level of precision consistently is an almost impossible task. At worst it leads to statements such as

> The interface to the system used by radar operators should be user friendly.
>
> The virtual interface shall be based on simple overall concepts which are straightforward to understand and use and which are few in number.

The former is at too high a level of abstraction and needs to be expanded to define requirements for help facilities, short versions of commands, and the text of user prompts. The latter is a platitude and should be removed from the statement of requirements.

Contradiction

A statement of requirements will often contain functional and non-functional requirements which are at variance with each other. In effect they eliminate the solution space of possible systems. Typically, the sentences that make up the contradictions will be scattered throughout the statement of requirements. An extreme example of such contradiction is the statement

> The water levels for the past three months should be stored on magnetic tape.

(which may form part of the hardware requirements of a future system) and the statement

> The command PRINT-LEVEL prints out the average water levels for a specified day during the past three months. The response of the system should be no longer than three seconds.

Obviously, if a slow-storage medium such as magnetic tape is used then the response time will hardly be in the range of a few seconds.

A more subtle error occurs with the statements

> Data is deposited into the employee file by means of the WRITE command. This command takes as parameters: the name of the employee, the employee's department, and salary.
>
> The ENTRY-CHECK command will print on the remote printer the name of each employee together with the date on which the employee's details were entered in the employee file.

which are functional requirements together with the non-functional requirement

> The hardware on which the system will be implemented consists of: an IBM PC with 512k store, asynchronous I/O ports, keyboard, monitor, and 20 Mb hard disc.

Here the assumption made is that the employee file will contain an entry date for each employee. Unfortunately, the WRITE command does not take an entry date as a parameter and the hardware specified does not include a description of a calender/clock.

A system cannot be developed which satisfies contradictory requirements. If this were regarded as a pure example of a contradiction, then the ENTRY-CHECK command should be deleted. However, the contradiction could have arisen from a set of incomplete requirements. In this case the WRITE command should be amended to take the entry date as a parameter or the hardware requirement expanded to include a calendar/clock.

Incompleteness

One of the most common faults in a statement of requirements is incompleteness. An example of this follows. It shows part of the functional requirements of a system to monitor chemical reactor temperatures.

> The system should maintain the hourly temperatures from sensors which are attached to functioning reactors. These values should be stored for the past three months.
>
> The function of the AVERAGE command is to display on a VDU the daily temperature of a reactor for a specified day.

These statements look correct. However, what happens if a user types in the AVERAGE command with a valid reactor name but for the current day? Should the system treat this as an error? Should it calculate the average temperature for the hours *up to* the hour during which the command was entered. Alternatively, should there be an hour threshold below which the command is treated as an error and, above which, the average temperature for the current day is displayed?

Mixed requirements

Rarely will you find functional requirements partitioned neatly into non-functional requirements, functional requirements, and data requirements. Often statements about a system's function are intermixed with statements about data that is to be processed.

Naivety

Another common failing of a customer statement of requirements is that it will contain naive views of what a computer system can achieve. This will be manifested in two ways. First, the statement of requirements will contain directives and statements which underestimate the power of the computer. The most frequent transgressors seem to be electronic engineers with little experience of software who insist on hardware requirements which could be easily satisfied by software at a much lower cost.

Another example of customer naivety occurs in statements of requirements for systems which can never be built within budget. Such systems are normally specified because of the low technical expertise of the customer. The most common example of requirements for an impossible system is the specification of a particular hardware configuration and a set of functions which will never meet its performance requirements.

Another example of naivety occurs when a customer suffers from a grossly ambitious view of what a system is capable of. One consequence of the recent rise in artificial intelligence has been a rash of statements of requirements which make the predictions of the wilder members of the artificial intelligence community seem almost sage-like.

Ambiguity

Statements of requirements written in natural language will almost always contain ambiguities. Natural language is an ideal medium for novels and poetry; indeed, its success depends on the large number of meanings that can be ascribed to a phrase or a sentence. However, it is a very poor medium for specifying a computer system with precision. Some examples of imprecision are

> The operator identity consists of the operator name and password; the password consists of six digits. It should be displayed on the security VDU and deposited in the login file when an operator logs into the system.
>
> When an error on a reactor overload is detected the *error1* screen should be displayed on the master console and the *error2* screen should be displayed on the link console with the header line continuously blinking.

In the first statement does the word 'it' refer to the password or the operator identity? In the second statement should both consoles display a blinking header line or should it only be displayed on the link console?

Mixtures of levels of abstraction

A statement of requirements will contain statements which are at different levels of detail. For example, the requirement

> The system should produce reports to management on the movement of all goods to and from all warehouses.

and the requirement

> The system should enable a manager to display, on a VDU, the cash value of all goods delivered from a specific warehouse on a particular day. The goods should be summarized into the categories described in section 2.6 of this document.

are at different levels of abstraction. The second requirement forms part of the first requirement. In a well-written statement of requirements the document should be organized into a hierarchy of paragraphs, subparagraphs, subsubparagraphs, etc. Each level of paragraph represents a refinement of the requirements embodied in the next higher level of paragraph. In a poorly written statement of requirements connected requirements will be spread randomly throughout the document.

2.4 The qualities of a good specification

The system specification is the document produced from requirements analysis. It represents the acceptable face of the statement of requirements and is the base document which is used by all subsequent developmental activities. As such, it should demonstrate properties which should be the opposite of those outlined in the previous section. For example, where the statement of requirements is vague and incomplete the system specification should be totally precise and include answers to every conceivable question the system designer would ask.

Rather than restate the opposites of the qualities described in the previous section, it shall be taken as read that a good specification should embody the opposite qualities described in that section. However, there are some additional qualities which the system specification should have.

Customer understandability

The document should be used as a medium of communication between the developer and the customer. The system specification is a document which completely describes the characteristics of a computer system as perceived by the developer. The only way to check that requirements analysis has been carried out correctly is by asking the customer to check the system specification. This checking can be carried out in a number of ways. First, the customer could read the specification and check it against the statement of requirements. Second, the developer could frame a number of questions for which tentative answers have already been prepared. These questions would be asked of the customer and the response checked with the tentative answers. Typical questions are

> What should happen if an intruder leaves primary radar space and enters secondary radar space?
>
> Should an invoice be rejected if it doesn't contain an order number?
>
> If the system is in the READY state and then is switched to the ACTIVE state while the data base is being updated what should happen to the recovery log?

Partitionability

The system specification should be partitioned. This should include both **vertical** and **horizontal** partitioning. Horizontal partitioning is the separation of the specification into functional requirements, non-functional requirements, data requirements, hardware requirements, and goals. It also means that each of these categories should be partitioned into logically equivalent sections. For example, all the functional requirements in a stock control system concerned with management reporting should be grouped together and separated from functional requirements for other parts of the system.

The major reason for horizontal partitioning is that it makes validation a much more manageable proposition. For example, if all the data requirements are grouped together, it is much easier to check for completeness and consistency than if they were scattered throughout hundreds of pages of a system specification.

Horizontal partitioning is also important in medium-to-large projects where one person from the developer's team cannot be expected to understand all of a proposed system. It allows customer staff with specialized skills to be involved in a well-defined part of the system specification. For example, a stores manager would only be interested in, and knowledgeable about, parts of a specification which concerned the movement of goods to and from his warehouse, and would not be knowledgeable about financial accounting functions which the company accountant would be interested in.

Vertical partitioning involves organizing a software specification into a series of levels. Each level corresponds to a level of abstraction in the statement of requirements this enables the specification to be examined by staff at different levels in the development team. For example, the designer in charge of overall system architecture may be interested only in the highest levels of detail, while staff engaged in specifying system tests may only be interested in the lower levels of the specification. A software specification should be written in such a way that it can be accessed by the developer's staff without having to worry about the intrusion of higher-level or lower-level concerns. Each chunk of the specification should be readable and understandable in isolation in the same way that individual program units in a coded software system can be understood in isolation.

A central question that should be asked by every software developer during requirements analysis is: how can I test that this requirement or set of requirements can be met by the developed software? No matter what notation is used for system specification—informal or formal—a developed system will have to undergo a series of acceptance and system tests to ensure that it meets user requirements. If the requirements embodied in the software specification are expressed in such a way that it is unclear whether an adequate test exists which validates these requirements, there is something drastically wrong. It usually means that the requirements are ambiguous or incomplete.

2.5 Carrying out analysis and system specification

Requirements analysis and specification consists of a series of stages. First, the customer statement of requirements should be examined in order to check that the proposed system is technically feasible, that is, that it can provide every function required subject to non-functional requirements and that the developer will make an adequate profit from the enterprise.

If the system is not feasible, the developer should negotiate a modified set of requirements. This will involve removing some of the functions of the proposed system or relaxing constraints such as the system performance, hardware configuration, or the cost of developing the system. If the customer agrees to the changes, the detailed requirements analysis can begin. If the customer disagrees, the project is cancelled. In a commercially tendered project this usually takes one of two forms. The first is to decline to make a bid. The second is to overbid.

Once the statement of requirements has been checked for feasibility, detailed requirements analysis and specification starts. This not only involves processing the customer statement of requirements, but also involves interviewing the customer's staff and examining any current manual and automated systems used by the customer. The major activity at this stage is to partition the requirements into functional, non-functional, hardware, and data requirements.

Each of these requirements should be specified as clearly as possible. Unfortunately, there are no adequate exact notations for non-functional and hardware requirements so natural language is invariably used. However, a number of exact notations are available for functional and data specification. The developer should select an appropriate notation. This book takes a view that discrete mathematics is the best one to employ for these parts of the specification.

The next stage is to identify any goals in the specification and rewrite them so that they are unambiguous and complete. Next, the design and implementation directives in the statement of requirements should be identified. It should be pointed out to the customer that by insisting on these he or she is reducing the probability of an optimal system being developed. The customer should then be asked to delete as many of these directives as possible from the statement of requirements, preferably all of them! If any remain, they should be noted in the system specification.

The final activity is to validate the system specification to check that it is an adequate reflection of the customer requirements. This validation is carried out jointly by the developer and the customer at a series of review meetings. It usually means the developer posing sets of questions representing stimulii and checking that the perceived response of the proposed system is not at variance with that of the customer.

One important activity which tends to get ignored during requirements analysis is the eliciting of future requirements from the customer. Software maintenance is a major developmental activity; a large part of it consists of modifying an existing system in order to respond to changing user requirements. We have now reached the position where we can design systems which are able to respond adequately to change. However, to be totally effective during the design process, the developer will need at least partial knowledge of future requirements. The original developer of a software system will normally be in a favourable position for a future maintenance contract. Thus, it is only prudent to identify future requirements so that a robust design can be developed.

If the developer is to elicit future requirements from a customer, the activity is best left to the very last stages of requirements analysis and specification. It is difficult enough to discover current requirements without having the dialogue between the developer and the customer cluttered up with a discussion of future growth.

2.6 An example of requirements analysis

Part of the statement of requirements for a system to monitor the state of a series of chemical reactors is shown below. It is reproduced for two reasons. First, to reinforce earlier comments on the inadequacy of natural language as a medium for specification. Second, in order to illustrate the type of questions that the specifier should be asking during the process of requirements analysis. These questions are described in the text following the statement of requirements.

> 1. The function of the monitoring system is to monitor and report on the operational state of the six chemical reactors at Ixburgh Nitrogen works.
>
> 2. Monitoring consists of checking that the temperature and pressure of each reactor does not exceed specified limits.
>
> 3. Reporting consists of allowing process workers to interrogate a stored file of temperature and pressure readings stored over the past 48 hours.
>
> 4. Readings of pressure and temperature are transferred along a serial line to a 486-based PC. Each reading consists of the chemical reactor number, the number of the sensor which originated the reading, a zero or one depending on whether the reading is a temperature or a

pressure, and, finally, the value of a reading. Each reading is issued every second.

5 On receiving a temperature or pressure reading that is outside safety limits the system should close down the offending reactor and display its name on the operator's console.

6. The system should allow plant operators to check on the state of operation of each reactor over the past 48 hours.

(a) The operator should be able to display on his or her VDU the average temperature or pressure from a specified sensor for any number of hours.

(b) The operator should be able to display on his or her VDU the number of sensor malfunctions for a specific sensor over any number of hours.

(c) The operator should be able to display on his or her VDU the name of those reactors which are currently functioning.

7. The operator interface should be as flexible as possible.

8. The system should display on the subsidiary console the current temperature and pressure of all functioning reactors for every sensor together with the average temperature and pressure over the past 60 seconds.

9. The system response should be the maximum possible achievable.

10. The system should be programmed in C. All file handling should be performed by calls on the operating system. The price of the contract will be £20 000. Legal details can be found in appendix 3.

This document is not far from the standard of many statements of requirements that I have met. A description of the deficiencies of each paragraph follows.

Paragraph 1

This states what the system is to do. It is a functional requirement. Moreover, it is a functional requirement at a high level of abstraction. At this stage it is worth noting the fact that six reactors are to be monitored and ask the customer about expansion plans later in the requirements analysis phase. If there is a high possibility of more reactors being built, it may be worth noting this down so that the system designer could allow for this in the size of the system files used to hold temperatures and pressure values.

Paragraph 2

Again this is a functional requirement. It is a little more specific than that described in paragraph 1. It mentions the fact that temperature and pressure should not exceed specified limits. Unfortunately, this paragraph does not specify what these limits are or refer to another part of the document which contains the limits. Also, how should the limits be communicated to the proposed system? Should they be part of the computer programs or should they be entered by the operator? If the reactors are to handle a wide variety of chemical processes, it will be worth asking the customer whether he wants an extra command to be implemented which would allow the operator to communicate the limits to the system.

Paragraph 3

This again is a functional requirement. It is at the same level of abstraction as that in paragraph 2. It also contains a design directive. The customer has asked for pressure and temperature values to be stored in the same file. Almost invariably, these values will be stored in backing storage but there is no reason why they cannot be stored in two separate files. Indeed, if they were stored in one file, it would involve the developer in unnecessary program complexity. The customer should be strongly advised to eliminate this requirement.

Paragraph 4

This is a data specification. It also mentions, as an aside, a hardware constraint. This data specification is incomplete: it does not give the size of each reading, it does not detail how many sensors there are, and does not specify whether a pressure reading will occupy the same number of bits as a temperature reading.

A large number of readings have to be stored in the temperature and pressure file and so the developer should check that the hardware requirements allow for this.

The paragraph also states that the readings of temperature and pressure will be transferred along *one* serial transmission line to the monitoring computer. If we assume that, say, two temperature and two pressure sensors are attached to each of the six reactors, will the pattern of readings transmitted along the line be of the form

Reactor 1 Temperature 1
Reactor 1 Temperature 2
Reactor 1 Pressure 1
Reactor 1 Pressure 2
...
...
Reactor 6 Pressure 1
Reactor 6 Pressure 2

This may seem a trivial question but monitoring equipment malfunctions, particularly in the harsh environment of a chemical reactor is vital. One manifestation of such a malfunction is an impossible reading such as a negative value or a missing reading. A later section of the requirements may ask for some function to be carried out if a malfunction occurs. Thus, it is vitally important to discover the pattern of readings in order to detect a malfunction.

Paragraph 5

Potentially this paragraph is the most dangerous in the whole statement of requirements. It could lead to massive increased costs on the project. In paragraph 1 it was stated that the function of the proposed system was monitoring and reporting. Paragraph 5 introduces a completely new major function: control. If this involves sending a simple signal to an existing and already programmed computer, this should be relatively straightforward. However, if the customer is also expecting the developer to produce the control software which causes the shut-down of a reactor, it introduces an added massive complexity into the system.

Another question that the developer should ask concerns the safety limits. One manifestation of a sensor malfunction is that it produces a reading that is outside safety limits. For example, a sensor which measures temperature between 20 and 1000° may produce an intermittent reading of 0° when it malfunctions. At this stage the software developer should establish what are reasonable values for readings and what values can be regarded as an indication of a malfunction.

This paragraph also contains the first reference to a reactor name. Previously in paragraphs 4 the reactor number had been referred to. It makes good sense for the interface to the system to be in terms of reactor names *and* sensor names. In which case how are these names to be associated with the corresponding reactor numbers and sensor numbers? Should the system dynamically allow the operator to assign and de-assign names and numbers? Finally, the paragraph mentions an 'operator'. Is this the same as the term 'process worker' in requirement 3?

Paragraph 6

This is a series of functional requirements which are an expansion of those requirements in paragraphs 3 and hence lie at a lower level of abstraction. Requirements 6(a) and 6(b) contain an ambiguity. It is associated with the phrase '... for a specific sensor for any number of hours'. Does this mean that an operator types in two hours and information about either temperature, pressure, or sensor malfunctions is displayed for the time between the two hours? This begs a further set of questions.

- What should the operator type if he wants to display information about temperature, pressure, or sensor malfunction for just one hour?

- What should happen if one of the hours typed is the current hour?
- What should be the format of the hours typed?

At first sight requirement 6(c) looks innocent. However, it mentions the fact that some reactors may not be functioning. If the reason for a non-functioning reactor is that it has been closed down because of a dangerous pressure or temperature reading, then this is consistent with paragraph 5. However, there may be other reasons for a non-functioning reactor. For example, a reactor may have closed down for routine maintenance. If other reasons exist, they may be hidden in later sections of the statements of requirements together with functions associated with closing down a reactor. There may even be a function concerned with starting up a reactor. Already paragraph 5 has alerted us to the fact that the proposed system may not be as simple as paragraph 1 had led us to believe.

This paragraph also contains the first reference to a sensor malfunction. No doubt a later paragraph will detail the function to be carried out when such a malfunction is detected.

Paragraph 7

This means nothing at all. The customer should be asked for a precise formulation. It possibly means that short versions of each command should be recognised by the system and that optional parameters should be catered for. It could also mean that some form of help facility should be provided. However, whatever the meaning, it should be made precise.

Paragraph 8

This is a functional requirement and is also a first reference to a subsidiary console. In paragraph 6 an operator's VDU was mentioned; it looks very much like two VDUs are required for the system. The final line of the paragraph also contains an ambiguity. Should the average pressure or temperature for the last 60 seconds be displayed as an average of all the sensors in a particular reactor? Perhaps the 60-second average should be displayed for each sensor?

Paragraph 9

This is a goal. Unfortunately it is not specified exactly as the words 'system response' are ambiguous. Do they mean the speed at which the system responds to operator queries or do they mean the speed with which a reactor can be closed down?

Paragraph 10

This consists of a series of constraints which reduce the possible solution space of system designs which implements the requirements of the system.

2.7 Mathematics and system specification

The previous section described in detail the painstaking approach that a software developer has to adopt in analysing customer requirements. Even with the small number of paragraphs shown, a large number of questions were asked and a large number of faults, ambiguities, and vague statements discovered. Many statements of requirements are much longer than 10 paragraphs of text. Consequently, the process of requirements analysis is the most difficult activity carried out by the software developer.

The system specification so produced should represent an amplification of the statement of requirements with all the faults removed. Because of the need to be precise such documents are often extremely bulky. The system specification for a recent American fighter plane occupied over 23 volumes of closely typed text. Theoretically, the system specification represents a correct and amplified version of the statement of requirements. What actually happens in practice is that, because of the nature of natural language and the size and complexity of current systems, many faults remain. In a sense, these faults are much more serious than those found in the statement of requirements: they tend to be spread out more in a bulkier document and are hence more difficult to detect during validation. In an attempt to alleviate many of the problems associated with requirements analysis and specification and the use of natural language, software engineers are beginning to turn to mathematics. Mathematics has a number of properties which are desirable for system specification.

It is easy to represent levels of abstraction

A simple example can demonstrate this. Assume that a mathematician wishes to differentiate the expression

$$\sin(x^2 + \exp(x^3 + 4))$$

He or she thinks of the expression as

$$\sin(u)$$

where u is $x^2 + \exp(x^3 + 4)$. Differentiating $\sin(u)$ gives

$$\cos(u)u'$$

The mathematician now has to differentiate u which is

$$x^2 + \exp(x^3 + 4)$$

to find u'. This can be thought of as

$$x^2 + \exp(w)$$

where w is $x^3 + 4$. Differentiating this gives

$$2x + w' \exp(w)$$

which is

$$2x + 3x^2 \exp(x^3 + 4)$$

Substituting back for u gives the differential as

$$\cos(x^2 + \exp(x^3 + 4))(2x + 3x^2 \exp(x^3 + 4))$$

In order to carry out the fairly complex operation of differentiating a large expression, the expression has been represented at various levels of abstraction. These are

$$\sin(u)$$

$$u = x^2 + \exp(w)$$

$$w = x^3 + 4$$

By doing this the mathematician has controlled the complexity of the problem by enabling the problem to be expressed as tractable pieces of mathematics.

As will be seen later in this book, the power of mathematics for representing levels of abstraction and for manipulating parts of expressions in isolation makes it an excellent medium for system specification.

It is easy to reason in mathematics

One of the characteristics of mathematics is its ability to deduce useful results, or to check results from propositions using rules of reasoning and theorems. Many of the readers of this book will remember learning geometric proofs which have chains of reasoning containing sentences such as:

> A triangle is a bounded figure which has three sides.
>
> If a figure is a triangle and the angle between two sides is a right angle, the triangle is a right-angled triangle.
>
> If a figure is a triangle and all the sides are equal, it is an equilateral triangle.
>
> If a line is drawn from the vertex of an equilateral triangle and this line bisects the opposite side, the two triangles so formed will be right-angled triangles.

There is no difference between these statements and those that are used when reasoning occurs during system specification. For example,

> If a reactor is shut down because of a malfunction or because of maintenance activity, the reactor can be regarded as closed down.
>
> If a reactor is closed down and an operator types the STATE command, a state error occurs.
>
> If a state error occurs, an error message will be displayed on the operator's console and a state error written to the error log file.

No mathematician would use natural language to represent such statements in a chain of reasoning and in justifying conclusions. Similarly, the need for precisely representing a specification and justifying chains of reasoning requires mathematics during the software project.

Mathematics is concise

Natural language is verbose. It contains a large number of noise words. Whenever long concepts are described or referred to they have to be written in full. Thus, terms such as 'the last currently malfunctioning reactor', 'the current ready state of the functioning reactors', and 'the last update transaction which has originated from a remote accounting location' may occur many times throughout a system specification. Even with paragraphs which express relatively straightforward functions the effect of such verbosity is to mask the essential details of a function.

More importantly, mathematics contains concepts which can be used to represent complicated relations that can only be expressed in a very large number of words.

Mathematics has well-defined and unambiguous meanings

Already this book has given a large number of examples of imprecision in natural language. Mathematics lies at the other end of the spectrum. A small and rather simple example of how mathematics is able to give more precision to a specification is shown below with the sentence

> The lamp voltage will always be a whole number of volts between 3 to 6 volts

and the mathematical expression

$$lamps \geq 3 \wedge lamps \leq 6$$

The former is ambiguous: does it mean that the voltages will include 3 volts and 6 volts? The mathematical equivalent answers this question as there is no possibility of a number of interpretations.

Mathematics can be used to model reality

One characteristic of an engineering process is the use, by its practitioners, of mathematics to model reality. The electronic engineer uses circuit theory and idealized models of elements such as resistors to design circuits; the civil engineer uses statics and dynamics to calculate stress and the peak loading of structures which have to carry traffic. In both cases mathematics is used to reason about a simplified but useful model of reality. If software engineering is to be regarded as a true engineering discipline, the same process ought to hold. The second part of this book describes a branch of mathematics which is used to model complex computer systems. The mathematics is taught in the context of describing real software systems. By the time that you have completed that section you should be convinced of the utility of mathematics in practical software development.

Mathematics can be used to suppress unnecessary detail

This occurs for a number of reasons. A very trivial one is the fact that a symbol can stand for a complex structure. A more important reason is the abstract properties of pure mathematics. Mathematicians are more interested in properties of objects rather than concrete values. Thus, a geometer is more interested in the fact that a square has equal sides rather than the values of the sides. Mathematics can be used to describe entities such as files, tables, and queues without worrying about concrete details such as the medium on which the entities are stored or the capacity of these structures, concerns which are not important during the functional specification process. A good specification notation should be sufficiently abstract to leave the system designer with a large solution space of possible designs, without providing concrete details which will emerge from the design process. Mathematics is the ideal medium for this.

Mathematics is constant

The mathematics taught in the next section of this book has been used for at least a century. It will always be taught in colleges, universities, and schools as it forms the basis of pure mathematics. We are currently afflicted with many new notations for specification and design many of which have proved transitory. Mathematics holds out the hope that it will provide a stable notation to describe the functionality and data of complex systems.

2.8 Problems with formal system specification

Before concluding this section it is only fair to point out some current problems and criticisms that have been voiced about formal methods of specification.

The customer cannot understand the software specification

A system specification which is expressed in mathematics can, at best, be understood only by the developer's staff. Clearly, it is impractical to expect an accountant or a middle manager from the customer's staff to understand pure mathematics.

The proponents of formal methods will say that in deriving a software specification expressed in mathematics the analyst must ask the customer a series of penetrating questions and that these questions are much more useful and lead to clearer specifications than if a natural-language description was used. This is true up to a point. Requirements analysis becomes a much more goal-oriented activity using mathematics; our limited experience of formal methods has shown that deeper questions about the properties of a system can be formulated leading to the discovery of major errors.

However, there is still a problem and it is this: a software specification is often a major reference document for the customer as well as the developer and often forms part of a legal contract; there must be a way of animating such a document so that the customer can understand it.

Very little work has been carried out in this area. One promising approach is to make a mathematical specification executable. This would enable the developer to prototype a system immediately after the system specification is complete.

There is a major technology transfer problem

This problem is more serious than the preceding one. While the mathematics of system specification is not particularly difficult, it is unfamiliar to the bulk of development staff. Even those staff who have studied mathematics at an advanced level in school will find the notation strange since many of our mathematics courses are still dominated by real-number mathematics such as calculus. There is evidence that the mathematics that is required for system specification is gradually percolating through to the schools; nevertheless, the vast majority of software developers regard it as unfamiliar and over-academic.

There is a second problem which is connected with technology transfer. Those books which describe formal methods still pay little attention to the mathematics. At best such books contain a mini-course of a few pages; at worst, they contain a reference to bone up on a particular pure mathematics book. This is understandable: the writer of a formal methods book wants to write a book on formal methods not a mathematics book. However, it does presuppose the existence of adequate preparatory material.

This book has been designed to alleviate the problems connected with such technology transfer. It teaches the mathematics in an informal and sympathetic way and shows how mathematics can be embodied in a particular specification notation known as Z. It would be easy to teach the mathematics involved in a fairly abstract way as most pre-computing texts do. This course has been rejected. The mathematics taught is always applied: whenever a new concept is introduced, it is illustrated with reference to requirements analysis and specification.

2.9 Further reading

[Davis, 1990] is a superb book which describes the process of requirements analysis in an incredible amount of detail. It contains a massive bibliography and will, I am sure, be the standard work on requirements analysis for many years to come. [Yourdon, 1989] is an extremely well-written description of the graphical approach to requirements analysis and system specification; this book is an object lesson in how to write clearly and concisely. [Mills, 1988] is a collection of readings from a pioneer of formal methods in the United States and gives one of the best rationales for the use of such methods that I know. [Meyer, 1985] is an excellent, short description of the advantages of mathematics over less formal notations. For readers who are interested in the nature of ambiguity [Empson, 1977] cannot be faulted. While abstraction is in general a good thing, sometimes an over-preoccupation with the idea can lead to unpleasant consequences. [Sacks, 1985] is a collection of neurological case studies which includes a description of a patient who was only able to perceive the world in terms of abstraction. I would recommend anyone interested in writing an accessible technical work to study this book: the author has an excellent style and is able to convey the most complex neurological ideas in a simple and effortless way. This book describes the formal notation Z. There are a number of other formal notations and development methods. Probably the most popular—apart from Z— is VDM. A good introduction to VDM can be found in [Andrews and Ince, 1991].

3

Propositional calculus

Aims

- To describe the nature of propositions and propositional expressions.
- To describe what mathematical proof involves.
- To describe how rules of inference can be used in the proof process.
- To show how propositional calculus can be used during requirements analysis and system specification.

3.1 Introduction

The process of system specification and requirements analysis involves examining the behaviour of a proposed software system and asking questions which enable a correct system specification to be developed. For example, the text of a statement of requirements may contain sentences such as

> If the reactor is on-line and the operator has typed in a query command, then the query log will be updated.
>
> If the operator has typed a query command from a secure terminal, then the command is allowed; otherwise, a security error will have occurred.
>
> If the system has been configured as being highly secure, then a message will be flashed on the master console when a security error occurs; otherwise, a violation message will be written to the validation log file.

The requirements analyst may take statements such as those above and formulate queries such as

> What events cause the query log to be updated?
>
> If a reactor is off-line and an operator types a query command then what will happen?

Table 3.1 The main propositional operators

Operator	Symbol
negation	$\neg$
conjunction	$\wedge$
disjunction	$\vee$
exclusive disjunction	$\vee_e$
implication	$\Rightarrow$
equality	$\Leftrightarrow$

In the example above the queries may need to be answered by examining fairly dense prose. What is required is a compact representation of the sentences, together with a means for enabling staff to reason about the proposed system. This is the role of the branch of pure mathematics known as **propositional calculus**.

3.2 Propositions and propositional operators

A proposition is a statement that can either be true or false. Typical examples from natural language specifications are

> The system is in shut-down mode.
>
> The ALARM command has been entered.
>
> An update file has been created.
>
> No sign-off messages have been received.

Such propositions can be represented as propositional variables which can be manipulated in the same way as algebraic variables. For example, the proposition 'the escape valve is open' can be represented by the propositional variable *EscValveOpen* and the proposition 'the reactor is in an error state' can be represented as the propositional variable *InErrorState*. Such propositional variables can have true values (true) or false values (false).

More complex propositions can be built up from propositional variables and a series of operators. These are shown in Table 3.1. Negation creates the opposite of a proposition. Thus, if *ValveOpen* stands for the proposition 'the escape valve is open' then $\neg$ *ValveOpen* stands for: 'the escape valve is closed'. The operator $\neg$ is an example of a **monadic** operator because it is applied to one argument.

The conjunction operator joins two propositions together to form another proposition; because of this it is known as a **dyadic** operator. The effect of the conjunction operator is to place the word *and* between its two arguments. Thus, if *ValveClosed* stands for 'the escape valve is closed' and *ShutDown* stands for 'the reactor has been shut down', then

$$ValveClosed \wedge ShutDown$$

stands for: 'the escape valve is closed and the reactor has been shut down'.

The disjunction operator is dyadic. The effect of this operator is to place the word *or* between its arguments. Thus, if *ValidCommand* stands for the proposition 'a valid command has been typed in at the console' and *error* stands for 'an operator error has occurred', then

$$ValidCommand \vee error$$

stands for the proposition 'a valid command has been typed in at the console or an operator error has occurred'.

The exclusive disjunction operator is dyadic. It is similar to the disjunction operator; however, it differs in one respect. It states that either one of its operands may be true but not both at the same time. Thus,

$$ValveClosed \vee ShutDown$$

asserts that either *ValveClosed* is true or *ShutDown* is true or that they are both true while

$$ValveClosed \vee_e ShutDown$$

asserts that either of the propositions are true, but not both at the same time.

The implication operator is dyadic. It describes the fact that one proposition implies another. The standard interpretation of the proposition

$$a \Rightarrow b$$

is 'if a then b'. For example, if the propositional variable *functioning* stands for 'the reactor is functioning normally', the propositional variable *ValidCommand* stands for 'a valid command has been typed', and the propositional variable *executed* stands for 'the command will have been executed', then

$$functioning \wedge ValidCommand \Rightarrow executed$$

stands for the proposition

> If the reactor is functioning normally and a valid command has been typed, then the command will have been executed.

The left-hand operand of the implication operator is known as the **antecedent** and right-hand operand is known as the **consequent**.

The final operator is the equality operator; again it is dyadic. It is similar to the equality operator of algebra. It is equivalent to the English phrases 'exactly when', 'only when', and 'if and only if'. Thus, if *ValveOpen* stands for 'the inlet valve is open', *MixerWorking* stands for 'the main mixer is working', and *NormalState* stands for 'the system is in a normal state', then the proposition

$$ValveOpen \wedge MixerWorking \Leftrightarrow NormalState$$

is equivalent to the proposition

> Only when the inlet valve is open and the main mixer is working will the system be in a normal state.

It is quite easy to confuse the meanings of implication and equality. An example, will make this clearer. The propositional expression

$$MainValveOpen \Rightarrow InletClosed$$

states that, when the main valve of a chemical reactor is open, the inlet to the reactor is closed, i.e. you can be sure that if you examine the main valve and it is open then it is certain that the inlet is closed. What it does not say is that if you examine the inlet and find it closed, you can be sure that the valve is open. The inlet may have been closed by an event which had nothing to do with the operation of the main valve; for example, a build up of sludge may have blocked the inlet. However, the propositional expression

$$MainValveOpen \Leftrightarrow InletClosed$$

states that *only when* the main valve is opened will the inlet be closed, i.e. if you examine the main valve and find it open, you can be sure that the inlet is closed and, if you examine the inlet and find it closed, you can be sure that the main valve is open.

Except for the implication operator the order of arguments of logical operators is irrelevant, for example,

$$ValveClosed \wedge SystemDown$$

is equivalent to

$$SystemDown \wedge ValveClosed$$

The names of the operators correspond to their meaning in English and large tracts of specifications can be converted into propositional calculus.

Worked example 3.1 A system specification contains the sentence

If the mixer has been activated and the inlet valve has been opened, the reactor is working normally.

Reduce this sentence to propositional calculus.

Solution The sentence contains propositions which can be expressed as the variables: *MixerActivated*, *ValveOpen*, and *ReactorNormal*. The sentence embodies an implication and can be written as

$$MixerActivated \land ValveOpen \Rightarrow ReactorNormal$$

The answer assumes, of course, that there may be other conditions under which the reactor will be working normally. If there were not, the implication operator would be replaced by the equality operator $\Leftrightarrow$
■

Worked example 3.2 Part of a system specification contains the sentences

1.1 There are only two valid commands which can be typed. They are the CHECKTRAFFIC command and the CHECKPOPULATION command.

1.2 Only if a valid command has been typed in by an authorized user will the command be executed normally.

1.3 If an invalid command has been typed and an authorized user has typed the command, then an invalid command error will be displayed.

1.4 If an unauthorized user types a command, then a warning screen (UNAUTHORIZED USER) is displayed on the master console and a violation error is written to the log file.

Reduce each sentence into propositional calculus form.

Solution The translation of each sentence is shown below.

$$CheckTraffic \lor CheckPopulation \Leftrightarrow ValidCommand$$

$$ValidCommand \land AuthorizedUser \Leftrightarrow NormalExecution$$

$$\neg\ ValidCommand \land AuthorizedUser \Rightarrow InvalidError$$

$$\neg\ AuthorizedUser \Leftrightarrow UnauthorizedScreen \land ViolationError$$

The first two propositions obviously involve the equality operator. The phrases 'only if' and 'only' make this clear. The fourth proposition also involves the equality operator since it is hard to see whether any other events would cause an unauthorized user screen to occur. However, in real life one would have to check the rest of the statement of requirements to ensure the correct use of the equality operator. The third proposition involves an implication operator as it is not clear from the fragment whether an invalid command error will occur when other events occur. For example, it is not quite clear whether an unauthorized user typing an invalid command will cause an invalid command error.

Note that it would have been just as correct to write the third proposition as

$$InvalidCommand \wedge AuthorizedUser \Rightarrow InvalidError$$

However, it would have introduced a superfluous propositional variable which would have added extra complexity to the system specification.
■

Exercise 3.1
Rewrite the following natural language sentences using propositional calculus. Assume that the $\Rightarrow$ operator will be used rather than the $\Leftrightarrow$ operator, unless, of course, the wording of a sentence explicitly makes equivalence clear—for example, by using the phrase 'if and only if'. Also assume that the $\vee$ operator is used unless the wording makes it clear that the $\vee_e$ operators is to be used.

(i) The reactor is in either of two states: the first state is that it is functioning normally; the second state is that it is shut down.
(ii) The user has typed a personal identifier and typed a group identifier.
(iii) Either the file will have been dumped to magnetic tape or it will be marked for archiving.
(iv) If the file has been marked for archiving and has not been used today then it will be archived.
(v) If a transmission line sends a ready signal and the receiver is on-line, then a link is started and a message is sent to the master console.
(vi) The receiver will only receive a ready message if the line is available.
(vii) If the update program has terminated with a transaction error, then a return message is sent to the operator and the program is aborted. If any program is aborted, then a termination message is sent to the current log file and the wasted time counter is updated.
(viii) If the file server is not on-line or the print server is not on-line, then the system will not be started.

■

So far the propositions which have been constructed have been relatively simple. However, when complex systems are specified, propositions will involve a number of operators and propositional variables. Typical expressions are

$$NotAvailable \wedge OutOfAction \vee InvalidTemp \Rightarrow NonFunctioning$$

and

$$\neg\ updated \lor \neg\ deleted \Rightarrow available \lor writing \lor reading$$

When complex expressions are written, there is a question about the order in which they are evaluated. This is the same problem that occurs in evaluating algebraic expressions in programming languages. The solution adopted in propositional calculus is exactly the same as that adopted for programming languages: it involves specifying the precedence of operators. The rules of precedence adopted for propositional calculus are:

1. Sequences of the same operators are evaluated from left to right, for example,

$$OldFile \lor NewFile \lor UpdatedFile$$

is equivalent to

$$((OldFile \lor NewFile) \lor UpdatedFile)$$

2. The order of evaluation of different, but adjacent, operators is given by the following order of precedence

$$\neg\ , \land, \lor, \Rightarrow, \Leftrightarrow$$

$\neg$ has the highest order of precedence and $\Leftrightarrow$ has the lowest precedence. Thus,

$$\neg\ active \land monitoring$$

is equivalent to

$$(\neg\ active) \land monitoring$$

and

$$OldFile \lor NewFile \Leftrightarrow ReadFile \land \neg\ WriteFile$$

is equivalent to

$$(OldFile \lor NewFile) \Leftrightarrow (ReadFile \land (\neg\ WriteFile))$$

Normally when complex propositional expressions are written it is often necessary to make wise use of brackets in order to make the meaning of the proposition clear.

Worked example 3.3 For the two propositions below indicate the order of evaluation by means of brackets

$$\neg\ OldFile \vee NewFile \Rightarrow DepositFile \Rightarrow NewEvent$$

$$new \wedge old \wedge modified \vee \neg\ updated \Rightarrow ValidEvent$$

Solution The first is

$$(((\neg\ OldFile) \vee NewFile) \Rightarrow DepositFile) \Rightarrow NewEvent$$

The second is

$$(((new \wedge old) \wedge modified) \vee (\neg\ updated)) \Rightarrow ValidEvent$$

Although the brackets are used to indicate the order of evaluation, it is usually unnecessary to write every possible set of brackets in order to make a proposition clear. For example, the first expression is best written as

$$((\neg\ OldFile \vee NewFile) \Rightarrow DepositFile) \Rightarrow NewEvent$$

and the second expression is best written as it stands.
■

Worked example 3.4 Enclose the following propositional expressions in brackets in order to make their meaning clear. Do not use superfluous brackets if you think they are not needed.

$$\neg\ ReadEvent \wedge OldEvent \Rightarrow ValidEvent \Rightarrow GlobalEvent$$

$$message \wedge deposited \Rightarrow \neg\ NewMessage \wedge OldMessage$$

Solution The first expression would be

$$(\neg\ ReadEvent \wedge OldEvent \Rightarrow ValidEvent) \Rightarrow GlobalEvent$$

Although bracketing is a personal thing I would say that no brackets are required in the second expression. The meaning is fairly clear from the propositional expression.
■

Exercise 3.2
Show the order of evaluation of the following expressions by using brackets.

Table 3.2 Truth table for the main propositional operators

a	b	$\neg a$	$a \wedge b$	$a \vee b$	$a \Rightarrow b$	$a \Leftrightarrow b$	$a \vee_e b$
false	false	true	false	false	true	true	false
false	true	true	false	true	true	false	true
true	false	false	false	true	false	false	true
true	true	false	true	true	true	true	false

(i) $\neg \neg$ *OldFileUpdated*
(ii) *NewReactorOnline* $\wedge$ *OldReactorOnline* $\vee$ *ErrorEvent*
(iii) *ValidCommand* $\wedge$ *operating* $\Rightarrow$ *ActionExecuted*
(iv) *TransmissionStarted* $\Rightarrow$ *LineOpen* $\Rightarrow$ *HandshakeOk*
(v) *EdCommand* $\vee$ *DelCommand* $\vee$ *DirectCommand*
(vi) $\neg$ *EdCommand* $\Leftrightarrow$ *DirectCommand* $\wedge$ *NormCommand* $\Rightarrow$ *OkState*

■

Up until now the syntax of propositions has been described. An informal semantics has been given in terms of natural language interpretations. However, the time has now arrived to give a precise meaning. This is embodied in a series of tables known as **truth tables**. Table 3.2 is a truth table for the operators described previously. The first two columns in Table 3.2 show possible values for the propositional variables a and b. Other columns indicate values when the propositional operators are applied. Thus, when a has the value false and b has the value false, $a \Rightarrow b$ has the value true and $a \vee b$ has the value false. Given the order of precedence of propositional operators, the value of propositional expressions can be calculated using Table 3.2. For example, the expression

$\neg$ *ValveOpen* $\wedge$ *malfunction* $\Rightarrow$ *ErrorReport* $\vee$ *ShutDown*

has the value true when *malfunction* has the value true, *ValveOpen* has the value true, *ErrorReport* has the value true, and *ShutDown* has the value false. The order of evaluation is

$\neg$ *ValveOpen* $\wedge$ *malfunction* $\Rightarrow$ *ErrorReport* $\vee$ *ShutDown*
$\neg$ true $\wedge$ *malfunction* $\Rightarrow$ *ErrorReport* $\vee$ *ShutDown*
false $\wedge$ *malfunction* $\Rightarrow$ *ErrorReport* $\vee$ *ShutDown*
false $\wedge$ true $\Rightarrow$ *ErrorReport* $\vee$ *ShutDown*
false $\Rightarrow$ *ErrorReport* $\vee$ *ShutDown*
false $\Rightarrow$ true $\vee$ *ShutDown*
false $\Rightarrow$ true $\vee$ false
false $\Rightarrow$ true
true

The evaluation of a fairly simple expression seems to have taken a large number of steps. However, after a large amount of practice, you will find that there are a number of shortcuts. For example, when a series of conjunctions

$$C_1 \wedge C_2 \wedge C_3 \wedge \ldots \wedge C_n$$

is encountered and one of the terms in the expression is false, then the whole expression is false. Similarly, if a series of disjunctions

$$C_1 \vee C_2 \vee C_3 \vee \ldots \vee C_n$$

is encountered and one of the terms is true, the whole expression is true. In evaluating an implication, if the antecedent has the value false, then the implication becomes true irrespective of the value of the consequent. Similarly, if the consequent of an implication has the value true then the implication is true irrespective of the value of the antecedent.

Exercise 3.3

What are the values of the propositional expressions shown below? Assume that the propositional variable *ValveOpen* has the value true, the variable *malfunction* has the value false, the variable *ValidCommand* has the value true and the variable *FlowOk* has the value true.

(i) $FlowOk \wedge ValveOpen \wedge malfunction$
(ii) $FlowOk \vee ValveOpen$
(iii) $\neg\, FlowOk \vee malfunction \vee ValveOpen$
(iv) $ValidCommand \wedge \neg\, malfunction \vee ValveOpen$
(v) $(ValidCommand \vee \neg\, ValveOpen) \wedge (FlowOk \Rightarrow ValveOpen)$
(vi) $\neg\,\neg\, FlowOk \Rightarrow malfunction \vee ValidCommand$
(vii) $ValveOpen \Rightarrow malfunction \Rightarrow ValidCommand$
(viii) $((FlowOk \Leftrightarrow \neg\, ValveOpen) \Rightarrow ValidCommand) \wedge \neg\, FlowOk$
(ix) $(\neg\,(malfunction \vee FlowOk)) \Rightarrow malfunction$
(x) $ValveOpen \Leftrightarrow malfunction \Leftrightarrow FlowOk \Rightarrow ValveOpen \Leftrightarrow FlowOk$
(xi) $\neg\,\neg\,\neg\,(FlowOk \Rightarrow (malfunction \wedge ValveOpen))$
(xii) $\neg\,(FlowOk \vee_e malfunction)$

■

3.3 Contradictions and tautologies

Two important types of propositions are known as **tautologies** and **contradictions**. A tautology is true for all possible values of the variables which make up the proposition. An example of a trivial tautology is

$$a \vee \neg\, a$$

Table 3.3 The truth table for a tautology

o	m	e	$(o \wedge m \wedge e) \Rightarrow (e \Rightarrow o)$
true	true	true	true
true	true	false	true
true	false	true	true
true	false	false	true
false	true	true	true
false	true	false	true
false	false	true	true
false	false	false	true

which is true for values of a of true and false. A trivial contradiction is

$$a \wedge \neg a$$

which is false for all values of a.

One way of discovering whether a propositional expression is a contradiction or a tautology is to evaluate it for all possible values of the variables that comprise the expression. For example, the proposition

$$(o \wedge m \wedge e) \Rightarrow (e \Rightarrow o)$$

can be demonstrated to be a tautology as shown in Table 3.3. For very large propositional expressions this method is clearly impractical. A later section will describe a more convenient way of discovering tautologies and contradictions.

3.4 Requirements analysis and specification

When requirements analysis and specification is carried out it involves a number of activities. First, the statement of requirements should be simplified. Second, any contradictions are removed. Third, any ambiguities should be made precise and the system specification validated with the customer. The system specification, so formed, should be precise and concise.

Propositional calculus by itself is not adequate for system specification. Later sections of this book will describe other notations which use the ideas presented in this chapter and which rely on propositional calculus as their logical basis. However, in order to give an early flavour of the utility of mathematics in system specification, this section will describe how propositional calculus can be used as a medium for the specification of relatively small and artificial systems.

First, expressing parts of a system specification in propositional calculus will always condense the specification. This form of simplification can be achieved by applying a series of rules known as **rules of inference** in the same way that rules can be used to simplify an expression such as

$$z = a(b + c)^2 - (ab^2 + ac^2 + abc)$$

to

$$abc$$

using rules such as

$$(x + y)^2 = x^2 + 2xy + y^2$$

Later in this chapter rules of inference for propositional calculus will be described.

Second, propositional calculus allows contradictions to be removed. For example, a system specification containing the sentence

> The system is in alert state only when it is waiting for an intruder and is on practice alert.

can be written as

$$alert \Leftrightarrow waiting \wedge practice$$

If a later paragraph contained the sentences

> If the system is in teaching mode and is on practice alert, then it is in alert state.
>
> The system will be in teaching mode and on practice alert and not waiting for an intruder.

then these propositions can be written as

$$teaching \wedge practice \Rightarrow alert$$

and

$$teaching \wedge practice \wedge \neg\, waiting$$

As will be seen later in this chapter, propositional calculus allows these expressions to be manipulated in order to show that a contradiction can be formed from the first propositional expression and the second and third propositional expressions and to deduce there was an error in the specification.

Now this is a simple example and could have been discovered by reading the natural language specification. However, in real statements of requirements, contradictions are almost invariably embedded in many layers of paragraphs where connections between propositions are invariably hidden in many levels of reference. In this case the use of a mathematical formalism is almost mandatory.

Third, propositional calculus enables ambiguities to be identified and removed. For example, consider the sentence

> If the inlet valve is open, then the system is switched to OPEN and the outlet valve is closed if the monitoring system is functioning.

How should this be written using propositional calculus? One interpretation is

$$InValveOpen \Rightarrow (MonSysFunctioning \Rightarrow open \wedge OutletClosed)$$

Another interpretation is

$$InValveOpen \Rightarrow (open \wedge (MonSysFunctioning \Rightarrow OutletClosed))$$

The fact that the analyst is forced into translating something imprecise into an exact notation enables hard questions about meaning to be posed. Such questions can be asked of the customer. The fact that the system specifier is forced to construct a mathematical document means that very close attention has to be paid to the natural language text of the statement of requirements.

Finally, a system specification expressed in propositional calculus enables the system specifier to frame questions to the customer in order to check that an interpretation of the statement of requirements is correct. For example, if a specification contained the proposition

$$AlertState \Rightarrow NormalAlert \vee practice$$

together with

$$NormalAlert \wedge MonitorTyped \Rightarrow error1$$

and

$$practice \wedge MonitorTyped \Rightarrow error1$$

a developer is able to frame questions such as

> If the system is in alert state and the operator typed a MONITOR command, would an error1 occur?

and

What events cause an error1?

Again this is a very simple example. Much more complicated examples will be described later. The purpose of this chapter is to give you the tools which will enable you to carry out requirements analysis and system specification using propositional calculus and other formalisms described in later chapters.

3.4.1 Simplification

Simplification of propositional expressions is carried out in the same way that algebraic expressions are simplified. For example, consider the statement involving the integer a

$$z = a * 1$$

This can be reduced to

$$z = a$$

from the fact that we know that if you multiply an integer by 1 the result is the integer that is multiplied. More formally we say that we infer that from $z = a * 1$ then $z = a$ follows. This can be embodied in something known as a **rule of inference**. The rule of inference corresponding to the simplification above is

$$\frac{z=x*1}{z=x} \qquad = mult1$$

The part above the line is a proposition which is true while the part below the line is another proposition which, given the truth of the first proposition, is also true. The remaining part of the rule of inference is its name $= mult1$. The equals symbol indicates that the rule works both ways: that if the top is true then the bottom is true and also that if the bottom is true then the top is true.

Another example of a law of inference involving three numbers a, b, and c, is shown below:

$$\frac{a>b, b>c}{a>c} \qquad trans >$$

It states that if a number a is greater than a number b and also b is greater than c then it follows that a is greater than c. Here the comma is used to separate two propositions which must be true. The fact that there is no $=$ operator in the name of the rule of inference indicates that the rule is one way: that if the top is true then the bottom is true, but not vice versa.

In order to simplify propositions and to check whether certain properties of a system specification hold a number of rules of inference are required. Many of these rules are described below.

The first two sets of rules involve the $\vee$ and $\wedge$ operators. The former are shown below. The symbols $p1$ and $p2$ stand for propositional expressions.

$$\frac{p1 \vee p1}{p1} = or1 \qquad \frac{p1 \vee \mathsf{true}}{\mathsf{true}} = or2$$

$$\frac{p1 \vee \mathsf{false}}{p1} = or3 \qquad \frac{p1 \vee (p1 \wedge p2)}{p1} = or4$$

$$\frac{p1}{p1 \vee p2} \quad or5$$

They allow certain propositional expressions involving the $\vee$ operator to be transformed to a simpler form. All these rules of inference can be derived from the truth table for each operator. For example, the rule $= or2$ can be derived from the fact that if you examine the truth table it can be seen that whatever the value of the first operand, if the second operand is true, then the value of the proposition is true.

The laws involving the $\wedge$ operator are shown below.

$$\frac{p1 \wedge p1}{p1} = and1 \qquad \frac{p1 \wedge \mathsf{true}}{p1} = and2$$

$$\frac{p1 \wedge \mathsf{false}}{\mathsf{false}} = and3 \qquad \frac{p1 \wedge (p1 \vee p2)}{p1} = and4$$

$$\frac{p1, p2}{p1 \wedge p2} = and5$$

There are a number of rules of inference involving the $\wedge$ and $\vee$ operators which describe the fact that these operators are commutative. That is, a proposition such as

The valve is open and the close command is typed.

is the same as the proposition

The close command is typed and the valve is open.

These are shown below.

$$\frac{p1 \wedge p2}{p2 \wedge p1} = comm \wedge \qquad \frac{p1 \vee p2}{p2 \vee p1} = comm \vee$$

There is also a rule of inference which describes the commutativity of $\Leftrightarrow$

$$\frac{p1 \Leftrightarrow p2}{p2 \Leftrightarrow p1} \quad = comm \Leftrightarrow$$

There are also a number of laws which are known as **associative laws**. These reflect the fact that brackets can be removed in propositional expressions involving the $\wedge$ and $\vee$ operators.

$$\frac{p1 \wedge p2 \wedge p3}{p1 \wedge (p2 \wedge p3)} \quad = rassoc \wedge \qquad \frac{p1 \wedge p2 \wedge p3}{(p1 \wedge p2) \wedge p3} \quad = lassoc \wedge$$

$$\frac{p1 \vee p2 \vee p3}{p1 \vee (p2 \vee p3)} \quad = rassoc \vee \qquad \frac{p1 \vee p2 \vee p3}{(p1 \vee p2) \vee p3} \quad = lassoc \vee$$

There are also a number of distributive laws which reflect the fact that propositions involving brackets can be expanded out.

$$\frac{p1 \vee (p2 \wedge p3)}{(p1 \vee p2) \wedge (p1 \vee p3)} \quad = dist \vee\wedge \qquad \frac{p1 \wedge (p2 \vee p3)}{(p1 \wedge p2) \vee (p1 \wedge p3)} \quad = dist \wedge\vee$$

An important set of rules of inference which were developed by Augustus De Morgan are shown below. They allow negation to be applied to the operands of the operators $\vee$ and $\wedge$. They are

$$\frac{\neg (p1 \vee p2)}{\neg p1 \wedge \neg p2} \quad = DeM \vee \qquad \frac{\neg (p1 \wedge p2)}{\neg p1 \vee \neg p2} \quad = DeM \wedge$$

The $= neg$ rule of inference allows double negatives to be eliminated

$$\frac{\neg \neg p1}{p1} \quad = neg$$

It states that a proposition such as

The reactor is not not functioning normally.

is equivalent to

The reactor is functioning normally.

The $= exmid$ rule of inference reflects the fact that a proposition can either be true *or* false. It is written as

$$\frac{p1 \vee \neg p1}{\mathsf{true}} \quad = exmid$$

The $= contr$ rule of inference reflects the fact that a proposition can neither be true or false.

$$\frac{p1 \wedge \neg\, p1}{\mathsf{false}} \quad = contr$$

There are a number of rules of inference which can be applied to implications; four are shown below.

$$\frac{p1 \Rightarrow p2}{\neg\, p1 \vee p2} \quad = impl \qquad \frac{p1, p1 \Rightarrow p2}{p2} \quad implies$$

$$\frac{p1 \Rightarrow p2, \neg\, p2}{\neg\, p1} \quad impliesnot$$

The $=$ *equal* rule of inference reflects the fact that equality can be removed from propositions which can then be written in terms of implication.

$$\frac{p1 \Leftrightarrow p2}{(p1 \Rightarrow p2) \wedge (p2 \Rightarrow p1)} \quad = equal$$

The $=$ *exor* law of inference allows exclusive or to be expressed in terms of or.

$$\frac{p1 \vee_e p2}{p1 \wedge \neg\, p2 \vee \neg\, p1 \wedge p2} \quad = exor$$

The *equiv* rule enables negations to be handled.

$$\frac{p1 \Leftrightarrow p2, \neg\, p2}{\neg\, p1} \quad equiv$$

These rules of inference hold no matter what the forms of $p1$, $p2$, and $p3$. For example, the expression

$$\neg \neg (open \vee closed \wedge operating)$$

can still be simplified to

$$(open \vee closed \wedge operating)$$

by means of the $=$ *neg* rule of inference. The main use of rules of inference is to simplify expressions, often to **true**, in which case a tautology has been found, or to **false**, in which case a contradiction has emerged. Often rules of inference are used to simplify a very complicated expression to one which can be more easily understood. A further use discussed later in this chapter is to support the reasoning process during requirements analysis.

As an example of the use of rules of inference consider the simplification of the expression

$$(open \vee closed) \wedge open$$

This can be transformed to the expression

$$open \wedge (open \vee closed)$$

by means of applying the $= comm \wedge$ law, and then transformed to the proposition

$$open$$

by application of the $= and4$ rule.

This can be expressed more formally as

$[tr1]$	$(open \vee closed) \wedge open$	
$[1]$	$open \wedge (open \vee closed)$	$[tr1, = comm \wedge]$
$[2]$	$open$	$[1, = and4]$

It is worth examining the structure of the table. The first proposition is labelled with a symbolic identifier, in this case *tr1*; the following lines which follow from the truth of this proposition are numbered sequentially. Each proposition which is numbered follows from a symbolically identified proposition which is assumed true or from a previously numbered line. So, for example, line 1 states that the proposition $(open \vee closed) \wedge open$ follows from the truth of the proposition $tr1$ by using the $= comm \wedge$ rule of inference. The part in square brackets gives the proposition which is assumed true followed by a comma and the rule of inference that was used. So, the form of display used in this book for proof consists of lines of true propositions which are either assumed true and labelled symbolically or which are derived to be true by means of applying rules of inference to earlier propositions which are true. Another example of this tableau-like exposition of proofs is shown below which describes the derivation of true from the propositional expression

$$((open \vee closed) \wedge malfunction) \vee \mathsf{true} \vee open$$

$[pr]$	$((open \vee closed) \wedge malfunction) \vee \mathsf{true} \vee open$	
$[1]$	$(open \wedge malfunction) \vee (closed \wedge malfunction) \vee \mathsf{true} \vee open$	$[pr, = comm \wedge, = dist \wedge\vee]$
$[2]$	$(open \wedge malfunction) \vee (closed \wedge malfunction) \vee \mathsf{true}$	$[1, = comm \vee, = or2]$
$[3]$	$(open \wedge malfunction) \vee (closed \vee \mathsf{true} \wedge malfunction \vee \mathsf{true})$	$[2, = comm \vee, = dist \vee\wedge]$
$[4]$	$(open \wedge malfunction) \vee (\mathsf{true} \wedge malfunction \vee \mathsf{true})$	$[3, = or2]$
$[5]$	$(open \wedge malfunction) \vee (\mathsf{true} \wedge \mathsf{true})$	$[4, = or2]$
$[6]$	$(open \wedge malfunction) \vee \mathsf{true}$	$[6, Defn \wedge]$
$[7]$	$(open \vee \mathsf{true}) \wedge (malfunction \vee \mathsf{true})$	$[6, = comm \vee, dist \vee\wedge]$

[8]	$(open \vee \mathsf{true}) \wedge \mathsf{true}$	$[7, = or2]$
[9]	$\mathsf{true} \wedge \mathsf{true}$	$[8, = or2]$
[10]	true	$[9, defn\ \wedge]$

There are two new aspects to this table that have not been discussed before. First, it often saves time and text to apply two or more rules of inference in one step. This is shown by writing the rules used separated by a comma. Second, occasionally when carrying out proofs there is a need to simplify an expression by recourse to the definition of the operators in the expression, for example, in line 10. This form of proof seems somewhat long-winded. However, at this stage in the book it is probably worth your while manipulating propositional expressions in short steps. When you have had more practice you can skip three or four steps and need not annotate your manipulations with 'obvious ' rules of inference such as those involving commutativity. In fact this example can be quickly transformed by recognizing that

$$(open \wedge malfunction) \vee (closed \wedge malfunction) \vee \mathsf{true} \vee open$$

is equivalent to $p1 \vee T$ where $p1$ is

$$(open \wedge malfunction) \vee (closed \wedge malfunction) \vee open$$

and this can be directly simplified to true by means of the $= or2$ rule of inference.

Worked example 3.5 Simplify the expression

$$\neg\neg(\neg((open \vee closed) \wedge open))$$

using the same form of display that was shown above.

Solution The display is shown below.

[exp]	$\neg\neg(\neg((open \vee closed) \wedge open))$	
[1]	$\neg\neg(\neg(open \wedge open \vee open \wedge closed))$	$[exp, = comm \wedge, = dist \wedge\vee]$
[2]	$\neg\neg(\neg(open \vee open \wedge closed))$	$[1, = and1]$
[3]	$\neg\neg(\neg\, open)$	$[2, = or4]$
[4]	$\neg\, open$	$[3, = neg]$

■

Exercise 3.4
Simplify the following propositional expressions; use the layout described in this section.

(i) $\mathsf{true} \wedge open$
(ii) $\mathsf{true} \wedge open \wedge \mathsf{false}$
(iii) $(\mathsf{true} \vee open) \wedge closed$
(iv) $malfunction \wedge open \wedge closed \wedge OperatorError \wedge \mathsf{false}$
(v) $(open \wedge \neg\, closed) \wedge open$
(vi) $(open \wedge \neg\, closed) \vee (open \wedge closed)$
(vii) $(open \vee open) \wedge (open \vee closed) \wedge open$
(viii) $\mathsf{true} \vee (open \Rightarrow closed \wedge malfunction) \vee open$
(ix) $\neg\neg\neg\, malfunction \wedge \neg\neg\neg\neg\, open$
(x) $\neg\neg\, (malfunction \vee open)$

■

We have seen how the rules of inference allow propositional expressions to be simplified. You will find that system specifications and statements of requirements will contain a large number of fragments where simplification can make the meaning of natural language statements clearer.

Consider the sentence taken from a system specification

> If the valve is opened, then the inlet controller is placed in a monitoring state, if the system is not in a startup mode.

This can be written as

$$ValveOpened \Rightarrow (\neg\, startup \Rightarrow monitoring)$$

which can be transformed using the $= impl$ rule of inference to

$$\neg\, ValveOpened \vee (\neg\, startup \Rightarrow monitoring)$$

which again can further be transformed by the $= impl$ rule of inference to

$$\neg\, ValveOpened \vee startup \vee monitoring$$

which stands for

> Either the valve is not opened or the system is in a startup mode or the inlet controller is in a monitoring state.

Although the implication has been removed from the 'simplified' expression it makes it clear that a disjunction of events occurs. This can be useful in examining propositions later in a system specification. For example, a later proposition may state that

> The valve is opened and the system is not in startup mode and the inlet controller is not in a monitoring state.

This can be written as

$$ValveOpened \wedge \neg\, startup \wedge \neg\, monitoring$$

But this contradicts the expression

$$\neg\, ValveOpened \vee startup \vee monitoring$$

derived above since there are no values which can be assigned to *ValveOpened*, *startup*, and *monitoring* which will make the expressions true.

Worked example 3.6 Express the following statements in propositional logic and derive the relationships between the monitoring state, the test state, and the fact that the reactor is operating.

> The reactor is either in the monitoring or test state. It cannot be in both states. The reactor is functioning normally only when it is in one of these states.
>
> If a reactor is functioning normally and is not in the test state then the reactor can be regarded as being operating.

There should be two propositions in your answer.

Solution The propositional expressions derived from the statements are

$$normal \Leftrightarrow monitoring \vee_e test$$

and

$$normal \wedge \neg\, test \Rightarrow operating$$

The first expression can be simplified by means of the = *exor* law to

$$normal \Leftrightarrow monitoring \wedge \neg\, test \vee test \wedge \neg\, monitoring$$

normal can then be substituted for in the second proposition giving

$$(monitoring \wedge \neg\, test \vee test \wedge \neg\, monitoring) \wedge \neg\, test \Rightarrow operating$$

which can be transformed to

$$monitoring \wedge \neg\, test \Rightarrow operating$$

by applying the $= \; dist \; \wedge\vee$, $= \; and1$, and $= \; contr$ laws of inference. This propositional expression could then be transformed into a form which doesn't involve an implication operator by applying the *impl* rule of inference. However, the form above is probably the simplest.

■

3.4.2 Reasoning

So far we have seen that the transformation from natural language into propositional calculus forces staff involved in specification to think more clearly and also enables then to simplify complicated statements. Another way that the transformation aids the analyst is with **reasoning**.

During the process of requirements analysis and specification development staff ask questions such as

> If the reactor is in the steady state and an update command is typed, does that mean that the reactor no longer remains in the steady state?
>
> If the database is empty and a query command is typed, does an error message get displayed?
>
> If a page break occurs at the same time as a chapter break, does the formatter eject a new page?

One aspect of reasoning involves the identification of a set of premises and a conclusion and then showing that the conclusion logically follows from the premises. Examples of premises which are taken from the questions above are

> reactor is in a steady state,
>
> update command is typed,
>
> database is empty,
>
> query command is typed,
>
> page break occurs,
>
> chapter break occurs.

Examples of conclusions taken from the questions above are

> reactor is not in a steady state,
>
> error message is displayed,
>
> formatter ejects a new page.

By examining premises and attempting to deduce conclusions development staff are able to check the completeness of a specification and discover inconsistencies.

The following is a fragment of a system specification from which premises can be extracted and conclusions deduced.

1. The purpose of the system is to check on the functioning of a series of chemical reactors.

2. Four commands are allowed. They are: TEMP, PRESSURE, INHIBIT and ERRORS.

3. The TEMP command has two parameters. The first is a temperature monitor number and the second is a chemical reactor number. When executed, this command will display the average temperature for the specified monitor in the specified reactor over a 12-hour period. If an error in a parameter occurs, then an ERRORTEMP message is displayed on the VDU and an ERRORTEMP log message is written to the log file.

4. The PRESSURE command has two parameters. The first is a pressure monitor number and the second is a reactor number. When executed, this command will display the average pressure for the specified monitor in the specified reactor over a 12-hour period. If an error in a parameter occurs, then an ERRORPRESSURE message is displayed on the VDU and an ERRORPRESSURE message is written to the log file.

5. The ERRORS command has two parameters. The first is a pressure monitor number or a temperature monitor number. When executed, this command will display the number of malfunctions that have occurred over the last 12 hours for a particular monitor attached to a specific reactor. If an error in a parameter occurs, then an ERRORMONITOR message is displayed on the VDU and an ERRORMONITOR message is written to the log file.

6. Writing an error message to the log file can be inhibited by means of the INHIBIT command. Such a command can only be typed by a privileged user or when the system is in test mode.

7. If the log file is not on-line, then, whenever the PRESSURE or TEMP command is typed, an OFFLINELOGFILE error is displayed. If an ERRORS command is typed then it will be executed normally.

Given this specification a number of questions can be asked involving premises and conclusions. A sample is shown below.

If *the system is in test mode* **(Premise)** will *the pressure command be executed normally?* **(Conclusion)**

If *the log file is off-line* and *the system is in test mode*, **(Premises)** will *the PRESSURE command be executed normally?* **(Conclusion)**

If *the temperature command is typed* with *the log file off-line* and *the operator mistypes both command parameters*, **(Premises)** will *an ERRORTEMP message be displayed on the VDU?* **(Conclusion)**

The conclusions to these premises can be shown to follow or not follow by means of a careful reading of the system specification. However, when such a specification occupies hundreds of pages then a degree of mathematical formalism is invariably required.

In most mathematical texts the relationship between premises and a conclusion is written in the form

$$P_1, P_2, P_3, \ldots, P_n \vdash C$$

It states that the conclusion C logically follows from the premises

$$P_1, P_2, P_3, \ldots, P_n$$

Thus, the fact that a conclusion *ValveOpen* follows from the premises *NormalState*, *OpenCommand*, and $\neg$ *TestState* can be written as

$$NormalState, OpenCommand, \neg\ TestState \vdash ValveOpen$$

It represents the question: would the valve be open if the reactor is in its normal state and the open command has been typed and the reactor is not in a test state? This is exactly the type of question that would be asked by a system specifier in order to clarify requirements.

Premises and conclusions can be simple propositions as above or can be formed from existing propositions and the propositional operators outlined in this chapter. For example,

$$open \wedge normal, \neg\ ErrState, open \Rightarrow TestState \wedge NormState \vdash open$$

Given a series of premises and conclusions, how can it be demonstrated that the conclusion follows from the premises? Two methods will be described in this section.

The first method involves transforming the premises and conclusions into a propositional form and simplifying by means of the laws of inference described earlier. The statement

$$P_1, P_2, P_3, \ldots, P_n \vdash C$$

can be written as

$$P_1 \wedge P_2 \wedge P_3 \wedge \ldots \wedge P_n \Rightarrow C$$

since the first statement asserts that, if P_1 and P_2 and $P_3, \ldots, P_n$ hold, then C holds. The $\wedge$ operator replaces the comma while the implication operator replaces the $\vdash$. Now, if it can be shown that

$$P_1 \wedge P_2 \wedge P_3 \wedge \ldots \wedge P_n \Rightarrow C$$

is always true, i.e. it is a tautology, then C has to be true (remember that if the antecedent of an implication is true and the implication is true, then the consequent is always true). The task is then to demonstrate that

$$P_1 \wedge P_2 \wedge P_3 \wedge \ldots \wedge P_n \Rightarrow C$$

is a tautology. This can be done using the rules of inference described previously.

As an example consider

$$\neg\, closed, open \vee closed \vdash open$$

which states that, if a valve is not closed and the valve is either open or closed, then it can be deduced that the valve is open. First, the expression is transformed to

$$\neg\, closed \wedge (open \vee closed) \Rightarrow open$$

This can be simplified as follows

$[pr]$	$\neg\, closed \wedge (open \vee closed) \Rightarrow open$	
[1]	$\neg\,(\neg\, closed \wedge (open \vee closed)) \vee open$	$[pr, = impl]$
[2]	$\neg\,\neg\, closed \vee \neg\,(open \vee closed) \vee open$	$[1, = DeM\ \wedge]$
[3]	$closed \vee \neg\,(open \vee closed) \vee open$	$[2, = neg]$
[4]	$closed \vee open \vee \neg\,(closed \vee open)$	$[3, = comm\ \vee]$
[5]	true	$[4, = exmid]$

Since the expression has been reduced to a tautology, it can be deduced that *open* can be inferred from $open \vee closed$ and $\neg\, closed$. Note that the final step of the simplification has telescoped a number of smaller steps. The expression: $closed \vee open \vee \neg\,(closed \vee open)$ has been identified as $p \vee \neg\, p$ where p is $closed \vee open$.

The stages described seem long and painstaking. However, by this time you should be able to telescope a number of the simplification steps. The main strategy which should be adopted is to eliminate the implication operator using the $= impl$ rule of inference, then use a combination of $= DeM$, $= dist$, $= or$, and $= and$ rules of inference to reduce the expression to true and hence demonstrate a tautology.

Worked example 3.7 Show that the conclusion $\neg\, LineClosed$ follows from the premises

$$ChannelOpen, ReceiverReady,$$
$$LineClosed \wedge ChannelOpen \Rightarrow \neg\, ReceiverReady$$

Solution It is necessary to demonstrate that

$$ChannelOpen \wedge ReceiverReady \wedge (LineClosed \wedge ChannelOpen \Rightarrow \neg\, ReceiverReady) \Rightarrow \neg\, LineClosed$$

First concentrate on simplifying the propositional expression to the left of the implication. The stages are

[*pro*]	$ChannelOpen \wedge ReceiverReady \wedge (LineClosed \wedge ChannelOpen \Rightarrow \neg\, ReceiverReady)$	
[1]	$ChannelOpen \wedge ReceiverReady \wedge (\neg\,(LineClosed \wedge ChannelOpen) \vee \neg\, ReceiverReady)$	$[pro, = impl]$
[2]	$ChannelOpen \wedge ReceiverReady \wedge (\neg\, LineClosed \vee \neg\, ChannelOpen \vee \neg\, ReceiverReady)$	$[1, = DeM\ \wedge]$
[3]	$\mathsf{false} \vee ChannelOpen \wedge ReceiverReady \wedge \neg\, LineClosed \vee \mathsf{false}$	$[2, = dist\ \wedge\vee, = contr]$
[4]	$ChannelOpen \wedge ReceiverReady \wedge \neg\, LineClosed$	$[3, = or3]$

To demonstrate the tautology it is thus necessary to show that

$$ChannelOpen \wedge ReceiverReady \wedge \neg\, LineClosed \Rightarrow \neg\, LineClosed$$

The sequence is

[*fro*]	$ChannelOpen \wedge ReceiverReady \wedge \neg\, LineClosed \Rightarrow \neg\, LineClosed$	
[1]	$\neg\,(ChannelOpen \wedge ReceiverReady) \vee LineClosed \vee \neg\, LineClosed$	$[fro, = impl, = DeM\ \wedge]$
[2]	$\neg\,(ChannelOpen \wedge ReceiverReady) \vee \mathsf{true}$	$[1, = exmid]$
[3]	true	$[2, = or2]$

■

Validating a specification involves attempting to show that an assumed conclusion can be derived from a series of premises generated from 'if questions', for example,

If the reactor is closed down and the computer malfunctions ...

If the file store is full ...

If the user types an invalid command ...

Almost invariably these premises alone will be inadequate to carry out the derivation. It will be necessary to use propositional expressions from the system specification in the deduction. If $P_1, P_2, \ldots, P_n$ are a series of propositions and C is a conclusion which is to be derived and $S_1, S_2, \ldots, S_m$ are a series of propositional expressions from the system specification, then it is almost always necessary to demonstrate that

$$P_1 \wedge P_2 \wedge \ldots \wedge P_n \wedge S_1 \wedge S_2 \wedge \ldots \wedge S_m \Rightarrow C$$

The propositional expressions $S_1, S_2, \ldots, S_m$ are usually those which involve variables which are contained in the premises and the conclusion to be inferred from the premises. This is illustrated in the next worked example.

Worked example 3.8 A system specification contains the sentences

1. If the operator types the ALARM command, then the subsidiary alarm will be activated and an alarm message will be written to the log file.
2. Whenever the subsidiary alarm is activated, the main valve is closed and shut-down is started.
3. If the main valve is closed then the reactor can be regarded as being in a non-operational state.

Express these sentences in propositional calculus. Also show that if the alarm command is activated the reactor will be in a non-operational state.

Solution The propositions are

$$alarm \Rightarrow subsid \wedge AlarmMessage$$
$$subsid \Rightarrow ValveClosed \wedge ShutDown$$
$$ValveClosed \Rightarrow NonOp$$

We shall label the above propositions S_1, S_2, and S_3. In order to demonstrate

$$alarm \vdash NonOp$$

it is necessary to demonstrate that

$$alarm \wedge S_1 \wedge S_2 \wedge S_3 \Rightarrow NonOp$$

This is shown as follows. First S_1 and S_2 are substituted in the predicate giving the first line of the proof.

$$[pr1] \qquad alarm \wedge (alarm \Rightarrow subsid \wedge AlarmMessage)$$

	$\wedge\ (subsid \Rightarrow ValveClosed \wedge ShutDown) \wedge S_3 \Rightarrow NonOp$	
[1]	$(alarm \wedge subsid \wedge AlarmMessage)$	
	$\wedge\ (subsid \Rightarrow ValveClosed \wedge ShutDown) \wedge S_3 \Rightarrow NonOp$	$[pr1, = imp, = dist \wedge\vee, = contr, = or3]$
[2]	$alarm \wedge subsid \wedge$	$[1, = impl, = contr, = and3, = or3]$
	$AlarmMessage \wedge ValveClosed \wedge ShutDown \wedge S_3 \Rightarrow NonOp$	
[3]	$alarm \wedge subsid \wedge AlarmMessage \wedge ValveClosed \wedge$	$[Substitution]$
	$ShutDown \wedge (ValveClosed \Rightarrow NonOp) \Rightarrow NonOp$	
[4]	$alarm \wedge subsid \wedge$	$[3, = impl, = dist \wedge\vee, = contr, = or3]$
	$AlarmMessage \wedge ShutDown \wedge ValveClosed \wedge NonOp \Rightarrow NonOp$	
[5]	$\neg\ (alarm \wedge subsid \wedge AlarmMessage \wedge ShutDown$	$[4, = impl]$
	$\wedge\ ValveClosed \wedge NonOp) \vee NonOp$	
[6]	$\neg\ (alarm \wedge subsid \wedge AlarmMessage \wedge ShutDown$	$[5, = DeM \wedge]$
	$\wedge\ ValveClosed) \vee \neg\ NonOp \vee NonOp$	
[7]	$\neg\ (alarm \wedge subsid \wedge AlarmMessage \wedge ShutDown \wedge$	$[6, = exmid]$
	$ValveClosed) \vee \mathsf{true}$	
[8]	true	$[7, = or2]$

■

Exercise 3.5
Show that the following hold.

(i) $open, open \vee steady \vdash open$
(ii) $test, functioning \vdash test \wedge functioning$
(iii) $receiving, receiving \Rightarrow open, open \vdash receiving \vee closed \Rightarrow open$
(iv) $open \wedge receiving \Rightarrow \neg\ closed, \neg\ open, receiving \Rightarrow closed \vdash$
$\neg\ (receiving \wedge open)$
(v) $a \Rightarrow \neg\ b, c \Rightarrow b, c \wedge d \vdash b$

■

Until now valid consequences have been shown to follow from a series of premises. Unfortunately, a system specification will always contain incomplete requirements, which give rise to premises which, when manipulated, do not lead to a desired conclusion. The example which follows consists of a series of paragraphs taken from different parts of a system specification.

> 1. Whenever a QUERY command is typed a query confirmation message is displayed and the system goes into query mode.

> 14. If a query confirmation message is displayed while the system is in training mode, then the originating VDU is locked out from incoming messages.
>
> 71. If an originating VDU is locked out from incoming·messages, then the system regards it as quiescent.

If the specifier wants to infer that when a query command is typed the system regards the originating VDU as quiescent, then

$$query \vdash quiescent$$

If the propositions from the specification are S_1, S_2, and S_3, where S_1 is

$$query \Rightarrow confirm \wedge QueryMode$$

S_2 is

$$confirm \wedge training \Rightarrow LockedVdu$$

and S_3 is

$$LockedVdu \Rightarrow quiescent$$

then it is necessary to demonstrate that

$$query \wedge S_1 \wedge S_2 \wedge S_3 \Rightarrow quiescent$$

is a tautology; however, when the expression is simplified using the rules of inference it gives

$$\begin{array}{l} \neg\,(query \wedge confirm \wedge QueryMode \wedge \neg\, training \wedge \\ \neg\, locked \vee query \wedge confirm \wedge QueryMode \wedge \\ \neg\, training \wedge quiescent \vee query \wedge confirm \wedge QueryMode \wedge locked \\ \wedge\, quiescent) \vee quiescent \end{array}$$

which cannot be simplified to true. Hence *quiescent* cannot be inferred from *query* and the propositional expressions from the system specification.

In this section I have described how by manipulating and simplifying propositional expressions a conclusion can be shown to be logically derived from a series of premises. This was achieved by applying rules of inference. Unfortunately, this technique can lead to a large amount of unwieldy and error-ridden manipulation, with large expressions being generated. An alternative, and quicker, technique is based on showing that the argument is invalid. It mimics the process whereby consequences are disputed in everyday discourse. If it is wished to demonstrate that

$$P_1, P_2, P_3, \ldots, P_n \vdash C$$

does not hold using a series of propositions $S_1, S_2, \ldots, S_m$ from a system specification, then all that is required is to show that when $P_1, P_2, P_3, \ldots, P_n$ and $S_1, S_2, \ldots, S_m$ have a **true** value then C has a **false** value. The way to proceed is to assume that C has a **false** value and that each P and S has a **true** value and analyse the consequences of an assignment of these values to the individual propositions that make up each P and S.

Such an analysis may lead to a contradiction; if it does, then this shows that C can be deduced from $P_1, P_2, P_3, \ldots, P_n$ using $S_1, S_2, \ldots, S_m$. However, if all the assumptions are satisfied and consistent then it has been demonstrated that C cannot be deduced from $P_1, P_2, P_3, \ldots, P_n$ and $S_1, S_2, \ldots, S_m$.

This is an example of another rule of inference. It is known as **indirect proof** or **proof by contradiction** or, in older mathematics textbooks, **reductio ad absurdum**.

An example will make this clearer. Assume that it is necessary to demonstrate

$$\neg\ ChannelOpen \vdash sender$$

and that $ChannelOpen \vee sender$ is known from the system specification. The proof is shown below

$[pr1]$	$\neg\ sender$	
$[pr2]$	$\neg\ ChannelOpen$	
$[pr3]$	$ChannelOpen \vee sender$	
$[1]$	$sender$	$[pr2, pr3, or3]$
$[2]$	$sender \wedge \neg\ sender$	$[pr1, 1, = and5]$
$[3]$	**false**	$[2, = contr]$

Each line is true and the final line derives the value **false** which shows a contradiction.

Another more complicated example is shown below. It involves the premises generated from the fragment of system specification

1. Whenever the system is in startup mode, all peripherals are closed down and, whenever the system is in functioning mode, then all peripherals are operating normally.
2. If all the peripherals are closed down and all peripherals are operating normally, then the system is in an inconsistent state.
3. The system can never be in an inconsistent state.

If the conclusion to be proved is that either the system is not in startup mode or not in functioning mode, then this can be written as

$$(startup \Rightarrow closed) \wedge (functioning \Rightarrow normal),$$
$$closed \wedge normal \Rightarrow inconsistent, \neg\, inconsistent \vdash$$
$$\neg\, startup \vee \neg\, functioning$$

The derivation of the conclusion is shown below

$[pr1]$	$(startup \Rightarrow closed) \wedge (functioning \Rightarrow normal)$	
$[pr2]$	$closed \wedge normal \Rightarrow inconsistent$	
$[pr3]$	$\neg\, inconsistent$	
$[pr4]$	$\neg\,(\neg\, startup \vee \neg\, functioning)$	
[1]	$startup \wedge functioning$	$[pr4, = DeM \vee, = neg]$
[2]	$startup$	$[1, = and5]$
[3]	$functioning$	$[1, = and5]$
[4]	$startup \Rightarrow closed$	$[pr1, = and5]$
[5]	$closed$	$[2, 4, implies]$
[6]	$functioning \Rightarrow normal$	$[pr1, = and5]$
[7]	$normal$	$[3, 6, implies]$
[8]	$inconsistent$	$[pr2, 5, 7, = and5, implies]$
[9]	false	$[pr3, 8, = contr]$

Worked example 3.9 By deriving a contradiction demonstrate that

$$open \Rightarrow normal \vdash normal \vee ShutDown$$

given the following two propositions from a system specification

$$closed \Rightarrow ShutDown$$

and

$$open \vee closed$$

Solution The proof is shown below.

$[pr1]$	$open \Rightarrow normal$
$[pr2]$	$closed \Rightarrow ShutDown$
$[pr3]$	$open \vee closed$
$[pr4]$	$\neg\,(normal \vee ShutDown)$

[1]	$\neg$ *normal* $\wedge \neg$ *ShutDown*	[$pr4, = DeM \vee$]
[2]	$\neg$ *normal*	[$1, = and5$]
[3]	$\neg$ *ShutDown*	[$1, = and5$]
[4]	$\neg$ *open*	[$pr1, 2, impliesnot$]
[5]	$\neg$ *closed*	[$pr2, 3, impliesnot$]
[6]	$\neg$ *open* $\wedge \neg$ *closed*	[$4, 5, = and5$]
[7]	false	[$pr3, 6, DeM \vee, = and5 = contr$]

■

So far indirect proof has been applied in demonstrating that a conclusion can be deduced from a set of premises in conjunction with a set of true propositions taken from a system specification. However, it can also be used to demonstrate that a conclusion cannot be deduced. An example of this is the demonstration that

$$scheduler \vdash \neg FirstFit$$

does not hold when taken in conjunction with the propositional expression

$$scheduler \Rightarrow FirstFit \vee BestFit$$

taken from a system specification.

[$pr1$]	*scheduler*	
[$pr2$]	*scheduler* $\Rightarrow$ *FirstFit* $\vee$ *BestFit*	
[$pr3$]	*FirstFit*	
[1]	*BestFit* $\vee$ *FirstFit*	[$pr3, or5$]
[2]	*scheduler* $\vee \neg$ *scheduler*	[$1, pr2, Defn \Rightarrow$]
[3]	**no contradiction**	

The important point to notice from this proof is that when line 3 is reached the propositions on the previous lines which are all true are unable to generate a contradiction.

Worked example 3.10 Demonstrate that the consequence

$$\neg LpAvailable$$

cannot follow from the premise

$$normal \Rightarrow printing$$

and the propositions

$$LpAvailable \vee spooler \Rightarrow normal \vee OnLine$$

and

$$\neg\ printing$$

taken from a system specification.

Solution The proof is shown below.

$[pr1]$	$LpAvailable \vee spooler \Rightarrow normal \vee OnLine$	
$[pr2]$	$normal \Rightarrow printing$	
$[pr3]$	$\neg\ printing$	
$[pr4]$	$LpAvailable$	
[1]	$\neg\ normal$	$[pr2, pr3, impliesnot]$
[2]	$LpAvailable \vee spooler$	$[pr4, or5]$
[3]	$normal \vee OnLine$	$[pr1, 2, implies]$
[4]	$OnLine$	$[1, 3, = or3]$
[5]	**no contradiction**	

■

Exercise 3.6
Demonstrate the validity or otherwise of the following inferences using indirect proof. In each question the inferences are written first followed by the propositions which are taken from a system specification.

(i) $old \vdash new$
$old \Rightarrow new$
(ii) $old \Rightarrow next \vdash new \Rightarrow next$
$old \Leftrightarrow new$
(iii) $\neg\ normal \vdash OffLine$
$OnLine \vee OffLine, OnLine \Rightarrow normal$
(iv) $directory \vee archive \Rightarrow normal, \neg\ normal \vdash \neg\ directory$
$on \vee directory$
(v) $\vdash LineDown \Rightarrow ready$
$LineOk \vee LineDown, LineOk \Rightarrow message \vee ready$

■

Table 3.4 A truth table

open	*closed*	*open* $\wedge$ *closed*	$\neg$ *open* $\vee$ $\neg$ *closed*
false	false	false	true
true	false	false	true
false	true	false	true
true	true	true	false

3.5 The detection of inconsistencies

Related to the problem of checking validity is the problem of determining the inconsistency of a set of statements which represent the premises for an inference. For example, the two propositions

$$ValveOpen \vee_e MonitorState$$

and

$$ValveOpen \Leftrightarrow MonitorState$$

are inconsistent since the first premise states that a valve can be open or the system can be in a monitor state but not both, but the second proposition states that if a valve is open then the system is in a monitor state and vice versa.

A series of propositional expressions are **consistent** if and only if there exists at least one assignment of truth values for the variables in each propositional expression such that all the expressions simultaneously receive the value true. For example, the propositional expressions

$$open \vee closed, \neg\, closed, closed \Rightarrow open$$

are consistent because there is one assignment to open (true) and closed (false) which leads to all the propositions having the value true.

Inconsistency is the reverse of this. A series of propositional expressions are **inconsistent** if and only if every assignment of truth values making up each propositional expression results in at least one expression receiving the value false. An example will make this clear. The propositional expressions

$$open \wedge closed, \neg\, open \vee \neg\, closed$$

are inconsistent. This can be seen from the truth table in Table 3.4. Whatever values are assigned to *open* and *closed* one of the propositional expressions will be false.

System specifications will contain inconsistencies and it would be useful if a systematic technique existed which could be used for their detection. Inconsistency can be detected in a way similar to that in which a conclusion can be deduced from a series of premises by the method of indirect proof. If by applying indirect proof to a series of propositional expressions we can generate a contradiction, then these statements are inconsistent. More formally, what is being attempted is a demonstration that

$$S_1, S_2, \ldots, S_n \vdash C$$

where C is any contradiction and $S_1, S_2, \ldots, S_n$ are a series of n propositional expressions taken from a system specification. As an example consider the propositional expressions

$$\textit{ready} \lor \textit{processing}, \neg\, \textit{ready} \land \neg\, \textit{processing}$$

If it is assumed that $\neg\, \textit{ready} \land \neg\, \textit{processing}$ is true, then *ready* must have the value false and also *processing* must have the value false. However, this contradicts the proposition $\textit{ready} \lor \textit{processing}$, since if it is true then at least one of its variables must be true. Since a contradiction has been generated the two statements are inconsistent.

This method can be formally justified as follows. Since it is intended to demonstrate that

$$S_1, S_2, \ldots, S_n \vdash C$$

This is equivalent to

$$S_1 \land S_2 \land \ldots \land S_n \Rightarrow C$$

Since C is a contradiction it must always have the value false. Thus, the above expression is equivalent to

$$S_1 \land S_2 \land \ldots \land S_n \Rightarrow \mathsf{false}$$

This only true if

$$S_1 \land S_2 \land \ldots \land S_n$$

is false. This conjunction will be false when at least one of its propositions is false. This corresponds to our informal definition of inconsistency.

The demonstration that a series of propositions is inconsistent can be laid out in the same way that the demonstration that a conclusion follows from a set of premises. Thus, the demonstration that the propositional expressions

$$\textit{ready} \lor \textit{processing}, \neg\, \textit{ready} \land \neg\, \textit{processing}$$

are inconsistent would be written as

$[pr1]$	$ready \lor processing$	
$[pr2]$	$\neg\, ready \land \neg\, processing$	
$[1]$	$\neg\,(ready \lor processing)$	$[pr2, = DeM\ \land]$
$[2]$	false	$[1, pr1, = and5, = contr]$

The only difference between a demonstration of inconsistency and the proof that a conclusion follows from a set of premises using indirect proof is that in the latter the conclusion is assigned the value false in advance, while in the former the final line is the contradiction.

A more complicated demonstration of inconsistency follows. It shows that the statements

$$open \Leftrightarrow normal, normal \Rightarrow functioning, \neg\, functioning \lor command,$$
$$\neg\, open \Rightarrow command, \neg\, command$$

are inconsistent.

$[pr1]$	$open \Leftrightarrow normal$	
$[pr2]$	$normal \Rightarrow functioning$	
$[pr3]$	$\neg\, functioning \lor command$	
$[pr4]$	$\neg\, open \Rightarrow command$	
$[pr5]$	$\neg\, command$	
$[1]$	$open$	$[pr4, pr5, impliesnot]$
$[2]$	$\neg\, functioning$	$[pr3, pr5, = or3]$
$[3]$	$\neg\, normal$	$[pr2, 2, impliesnot]$
$[4]$	$\neg\, open$	$[pr1, 3, equiv]$
$[5]$	false	$[1, 4, = and5, = contr]$

Worked example 3.11 A system specification contains the statements shown below. Determine whether they are inconsistent.

1. If the system is in update mode, then it is always functioning normally and vice versa.
2. If the system is in update mode, then all the incoming messages will be queued.
3. If all incoming messages are not queued, then they will be diverted to the overflow file.
4. If the system is not in update mode, then incoming messages will be diverted to the overflow file.
5. Incoming messages will never be diverted to the overflow file.

Use the techniques outlined in this chapter.

Solution The sentences can be written as

$$update \Leftrightarrow functioning, update \Rightarrow queued, \neg\, queued \Rightarrow diverted$$
$$\neg\, update \Rightarrow diverted, \neg\, diverted$$

The attempted proof to demonstrate inconsistency is as follows.

$[pr1]$	$update \Leftrightarrow functioning$	
$[pr2]$	$update \Rightarrow queued$	
$[pr3]$	$\neg\, queued \Rightarrow diverted$	
$[pr4]$	$\neg\, update \Rightarrow diverted$	
$[pr5]$	$\neg\, diverted$	
[1]	$update$	$[pr4, pr5, impliesnot]$
[2]	$queued$	$[1, pr2, implies]$
[3]	$functioning$	$[pr1, 1, equiv]$

No contradiction can be generated; hence the statements are consistent.
∎

4

Predicate calculus

Aims

- To show how predicate calculus can represent statements about collections of objects.
- To describe some rules of inference in the predicate calculus.
- To show how predicate calculus can be used for program specification.

4.1 Introduction

In the previous chapter propositional calculus was described. System specifications and statements of requirements were used as concrete examples of its utility. It was stressed that the examples used were artificial in the sense that they were small. However, the examples were artificial in another respect. They were carefully chosen so that they could be expressed in propositional calculus. Unfortunately, the propositional calculus alone is an inadequate medium for expressing many system specifications. As an example consider the two propositions

> All the monitoring computers will be ready.

and

> X12 is a monitoring computer.

Using propositional calculus it is not possible to prove that

> X12 will not be ready.

And from the propositions

> If a thermocouple registers a temperature outside the range 10 to 100 degrees centigrade then the thermocouple is malfunctioning.

and

> ThermocoupleA is reading 112 degree centigrade.

it is not possible to prove that

> ThermocoupleA is malfunctioning.

These examples demonstrate that propositional calculus has two weaknesses. The first example demonstrates that it is not possible to reason about classes of objects. Thus, sentences such as

> At least one user is logged on.
>
> All thermocouples will either be functioning normally or malfunctioning.
>
> If all the valves are open, then at least one monitoring computer will be on-line.

cannot be adequately expressed. The second example demonstrates that propositional calculus manipulates objects which are not fine-grained enough to reason adequately about. For example the sentence

> If one of the line voltages lies outside the range 110 to 200 degrees centigrade, then the line is malfunctioning.

could be represented as the propositional expression

$$LineVolt110to200 \Rightarrow malfunction$$

but such a proposition is inadequate to reason with, if for example, later sentences referred to temperature values. What is required is a means of splitting up propositions such as $LineVolt110to200$ into its constituent parts.

4.2 Propositions as predicates

In order to overcome some of the disadvantages of propositional calculus the idea of a proposition will be generalized in two ways. The first generalization is that a proposition can be replaced by any expression which can be evaluated to either true or false. This chapter will limit these expressions to contain: identifiers which have integer values, integer constants, relational operators applied to integer variables and constants, and arithmetic operators applied to integers.

Such expressions will be written in the same way that they would be written in a computer program with the occasional use of brackets to make their meaning clear. Thus,

$$NewTemp > 10$$

$$(OldTemp * 10 + NewTemp) > MaxLimit$$

$$OldMonitor > 10 \Rightarrow NewMonitor > 10$$

are all examples of expressions which can replace propositions.

Propositions, as described in this book, can also include a construct which describes the relationship between two objects which can be variables or constants. These objects can be of any type; for example, they may be computers, users, files, or monitoring instruments. Relationships are written in the form

$$name(variable\ or\ constant, variable\ or\ constant)$$

At this stage of the book such a construct can be regarded as being similar to a boolean function which has two parameters of an arbitrary type. An example of the construct is

$$IsConnected(computer1, computer7)$$

which will return a value true when *computer*1 is connected to *computer*7 and false otherwise, and

$$FileState(file1, deleted)$$

returns a value true when the file *file*1 has been deleted and false otherwise. The objects that can replace propositions in the propositional calculus are known as **predicates**. The combined use of predicates and the operators of propositional calculus gives rise to the **predicate calculus**.

Some examples of statements taken from system specifications together with their predicate calculus equivalents follow.

The reactor temperature lies between 0 and 10 degrees centigrade.	$temp \geq 0 \land temp \leq 10$
Operator commands will always consist of three letters.	$CommandLength(command, 3)$
If the line voltage is active, then the line is malfunctioning.	$LineVoltage(line, active) \Rightarrow LineState(line, malfunctioning)$
If a line is malfunctioning, then the line has been disconnected.	$LineState(line, malfunctioning) \Rightarrow connection(line, disconnected)$
Whenever the computer is brought	$ComputerState(computer, offline) \Rightarrow$

off-line, the communications line is set to a high value of 5 volts.	$ComVolt = 5$
If the temperature of the main reactor is over 1000 degrees, then the monitoring computer is placed in an alarm state.	$temp(MainReactor, MainReacTemp) \wedge MainReacTemp > 1000 \Rightarrow CompState(MonitoringComp, alarm)$

The use of *MainReacTemp* in the last predicate often gives some trouble to the reader. This states that if the temperature of the main reactor is *MainReacTemp* and if its value is over 1000 degrees centigrade, then ...

Worked example 4.1 Convert the following extract from a natural language specification into predicate calculus form.

> Only when the main computer is in a monitoring state will the subsidiary computer be in a monitoring state.

Assume that implication is not used.

Solution The predicate employs the $\Leftrightarrow$ operator as it is clear that the only time that the subsidiary computer is in the monitoring state is when the main computer is in a monitoring state. The predicate will then be

$$CompState(main, monitoring) \Leftrightarrow CompState(subsid, monitoring)$$

■

Exercise 4.1

Convert the following sentences into predicate calculus form.

(i) A transaction is invalid if its amount is greater than £500.
(ii) A transaction is valid if the account number is greater than 1000 and the transaction amount lies between £10 and £200.
(iii) Whenever a sensor malfunctions then the master console is activated and the sensor voltage is grounded at 0 volts.
(iv) There will never be more than seven reactors on-line.
(v) If more than six VDUs are currently active, and provided that the second processor is on-line, then a first-fit algorithm will be used for storage allocation.
(vi) If more than three quarters of memory is occupied by programs, then no more than 12 interactive users will be allowed on the system at one time.
(vii) A transaction will never contain more than five parameters and never have more than 20 trailing blocks and be greater than 150

characters in length.
(viii) If the length of the address field is more than 10 characters then the transaction can be regarded as rejected.

∎

Exercise 4.2
Translate the following statements expressed in predicate calculus into English.

(i) $CompState(main, CompDown) \Rightarrow LineVolts = 0$
(ii) $NoUsers < 10 \wedge NoTerminals < 15$
(iii) $AmbientTemp > 50 \vee \neg\, AmbientTemp > 100 \Rightarrow$ $LineVolts = 40$
(iv) $PeriphState(lprinter, busy) \vee PeriphState(lprinter, down) \Rightarrow$ $\neg\, connected(lprinter, printbus)$
(v) $\neg\,(NoUsers \geq 10 \wedge NoFiles \leq 100)$
(vi) $MalfunctioningSensors > SensorNumbers * 2 \Rightarrow$ $state(MainComputer, idle)$
(vii) $\neg\, PeriphState(transducer, malfunctioning) \Rightarrow$ $CompState(main, idle) \vee IsConnected(main, subsid)$

∎

Exercise 4.3
The following is an extract from the specification of a financial system. Convert it to predicate calculus.

1. The system is to process a number of transactions stored in an update file. There will never be more than 100 transactions in the file.
2. There are two types of transactions: account transactions and report transactions. The account transaction will never contain more than five fields unless the first field contains an integer 1; in this case the transaction will contain between five and 10 fields. The report transaction will contain either two or three fields.
3. If the update file contains more than 80 transactions, then a *warning1* message should be sent to the operator's VDU. If the update file contains more than 90 transactions, then a *warning2* message should be sent to the operator's VDU.
4. If a transaction contains more than the allowable number of fields, then a *warning3* message is sent to the operator's VDU.

∎

4.3 Quantifiers as predicates

Propositional calculus is unable to adequately express statements about the properties of classes of objects and to reason about them. For example, given the propositions

> If one of the sensors is malfunctioning, then a zero reading is received at the communications terminal.

> Whenever a zero reading is received at the communications terminal, a process is allocated to deal with the malfunction provided that the process is in a ready state.

We are unable to deduce that, if SE11 is a sensor and is malfunctioning, then the malfunction will be dealt with by allocating a ready process.

What is required is a mathematical shorthand for expressing propositions that hold for a class of objects. The mathematical shorthand involves two operators known as the **existential quantifier** and the **universal quantifier**. The former is written as a mirror image of the capital letter E ($\exists$). The latter is written as an inverted capital A ($\forall$).

4.3.1 Existential quantification

The existential operator asserts that at least one object in a collection of objects has a particular property. Two examples of the use of the existential quantifier are

$$\exists\, i : 1 \,..\, 10 \bullet i^2 = 64$$

and

$$\exists\, proc : processors \bullet ProcessorState(proc, active)$$

The first example asserts that there is a value of i in $1, 2, 3, \ldots, 10$ which has a square equal to 64. The second example asserts that there is a processor which is currently active. The general form for existential quantification is

$$\exists\, identifier/classlist \bullet predicate$$

where *identifier/classlist* is a series of identifiers separated by commas followed by a colon and a class. Each *identifier/classlist* is separated by semicolons.

Class represents a class or set of objects over which the quantification holds. At this stage in the book *class* will be restricted to a range expression consisting of two integers separated by two periods; for example,

$$7 \,..\, 26$$

is the set of numbers $7, 8, 9, \ldots, 26$, the natural numbers $\mathbb{N}, (0, 1, 2, \ldots)$, the integers $(-\infty, \ldots, -1, 0, +1, \ldots, \infty)$, and sets of objects such as *processors*, *TransactionTypes*, or *Monitors*. Examples of the use of $\exists$ follow.

$$\exists\, i : \mathbb{N} \bullet i > 10 \land MonitorTemp = i \qquad (1)$$
$$\exists\, m : AllocatedMonitors \bullet MonState(m, ready) \qquad (2)$$
$$\exists\, t : transactions \bullet TransState(t, valid) \qquad (3)$$

$$\exists\, i : 1\,..\,100;\ m : AllocatedMonitors \bullet \quad (4)$$
$$\quad activity(m, functioning) \land AmbientTemp = i$$
$$\exists\, r : CurrentReactors;\ m : AllocatedMonitors \bullet \quad (5)$$
$$\quad MonState(m, functioning) \land connected(r, m)$$
$$\exists\, vdu : AllocatedVdus \bullet VduStatus(vdu, on) \Rightarrow \quad (6)$$
$$\quad GlobalStatus(vdu, active)$$

(1) is a long-winded way of saying that *MonitorTemp* is greater than 10. (2) states that out of all allocated monitors at least one is ready. (3) states that out of all transactions at least one is valid. (4) states that out of all allocated monitors there will be at least one which is functioning when the ambient temperature is between 1 and 100 degrees centigrade. (5) states that there is at least one functioning monitor which is connected to a reactor. (6) states that there is at least one allocated vdu which, when it is in the on state, will be regarded as active.

Worked example 4.2 Convert the following sentence into predicate calculus form.

> One of the actuators will be functioning provided that the system ambient temperature is within safety limits and the main reactor is not in a warm-up state. The current safety limits are 10 to 100 degrees centigrade.

Assume that the system ambient temperature is represented by the integer variable *AmbTemperature*.

Solution The solution is

$$AmbTemperature \geq 10 \land AmbTemperature \leq 100 \land$$
$$\neg\ ReactorState(main, WarmUp) \Rightarrow$$
$$\exists\, a : acts \bullet ActuatorState(a, functioning)$$

∎

Worked example 4.3 Convert the following sentences into predicate calculus form.

> 1. The queueing system handles the processing of retrieval messages from remote systems. There will be a maximum of 25 queues.
> 2. One of the queues will always be in an active state.

3. At least one queue will be in an active state and awaiting messages from priority users.

4. No queue will ever be in an active state, awaiting messages from priority users and in a ready state.

Assume that the number of queues is represented by the integer variable *NoQueues*; *QueueStatus*(,) is a predicate which gives the status of a queue; and *WaitStatus*(,) is a predicate which describes whether a queue is waiting for messages from remote users. *AvailableQueues* contains the name of the queues.

Solution The predicates are

$$NoQueues \leq 25$$

$$\exists\, queue : AvailableQueues \bullet \\ \quad QueueStatus(queue, active)$$

$$\exists\, queue : AvailableQueues \bullet \\ \quad QueueStatus(queue, active) \wedge WaitStatus(queue, waiting)$$

$$\neg\ \exists\, queue : AvailableQueues \bullet \\ \quad QueueStatus(queue, active) \wedge WaitStatus(queue, waiting) \\ \quad \wedge\ QueueStatus(queue, ready)$$

■

Exercise 4.4
Convert the following sentences into predicate calculus form.

(i) At least one command line will be active.
(ii) No files in the file store will be marked for archiving and marked for deletion.
(iii) At least one transaction will be a delete transaction provided that the transaction file is in a ready state.
(iv) No directories in the file store will be open or no file in the file store will be closed.
(v) At least one VDU will be connected to the peripheral processor.
(vi) If the main valve is open and one of the exhaust valves is open, then the control signal will be in a ready state providing that the ambient reactor pressure lies between 0 and 100 mm.
(vii) Either at least one communication line will be active or at least one

response line will be active.

(viii) If at least one communication line is down, then the system is switched to a warning state.

(ix) If at least one transaction is a delete transaction or an update , transaction then at least one update record will be modified.

■

There are a number of important features of quantifiers. First, an expression which is quantified is itself a predicate and thus can be used in conjunction with other predicates. Thus, predicates such as

$$\neg \exists program : FileStore \bullet name(program, tracker) \lor$$
$$\exists file : FileStore \bullet \neg FileState(file, deleted) \lor$$
$$\exists arch : ArchiveStore \bullet \neg ArchiveState(arch, deleted)$$

can be formed from quantifiers and the standard propositional operators such as $\land$, $\lor$, and $\neg$. Second, since quantified expressions are predicates, they can have a value true or false. For example,

$$\exists i : 1 \ldots 10 \bullet i^2 = 144$$

is patently false since there is no number between 1 and 10 which has a square of 144, while

$$\exists j : 1 \ldots 100 \bullet j \geq 3 \land j \leq 8 \land j^2 = 49$$

is true ($j = 7$). Third, quantification involves a different use of variables from that found in the propositional calculus. As an example consider the predicate

$$\exists j : 1 \ldots 50 \bullet j = MinTemp$$

This is equivalent to the predicate

$$MinTemp \geq 1 \land MinTemp \leq 50$$

Now in the first predicate j takes no part. Obviously, identifier j takes a different role from that of $MinTemp$.

Identifier $MinTemp$ is known as a **free variable** while j is known as a **bound variable**. Bound variables use scoping rules similar to those used in languages such as Pascal and Ada. Any bound variables are recognized as such within the predicate in which is it quantified; an identifier identical to a bound variable, but used outside a quantification is regarded as a free variable. Thus, in

$$CompState(m, active) \lor$$
$$\exists m : WorkingMonitors \bullet MonState(m, ready)$$

the first occurrence of m is an example of it being used as a free variable; the second occurrence is an example of it being used as a bound variable.

4.3.2 Universal quantification

So far this section has shown how statements which assert the existence of *one* object can be written using existential quantification. There is also a need to express the fact that *all* objects in a certain class have a particular property. Typical expressions of this taken from system specifications and statements of requirements are shown below.

> Unless the exhaust valve is closed, all the inlet valves will be open and all the transfer valves will be closed.
>
> All operator commands will be received from either the master console or, in the case of an emergency, from the subsidiary console.
>
> A program in main memory will either be suspended, ready, or active.

The first two statements are obviously connected with asserting a property of a class of objects. The first concerns the class of inlet valves and transfer valves; the second concerns the class of operator commands. The third statement asserts a property of objects in a class, but not obviously so. If the statement is written in the form: 'Every program in main memory ...' it is much clearer.

To handle such statements a second quantifier is used. It is known as the **universal quantifier** and is written as an inverted capital A ($\forall$). It has the same form as the existential quantifier. Some examples of its uses are

$$\forall i : 1 \,..\, 10 \bullet i^3 \leq 1000$$

$$\forall \, valve : ExhaustValves \bullet \\ \quad ValveStatus(valve, open) \vee_e ValveStatus(valve, closed)$$

$$\forall \, prog : InMemoryProgs \bullet \\ \quad ProgStatus(prog, suspended) \wedge \neg \, ProgStatus(prog, active)$$

The first example states that all the integers between 1 and 10 have their cubes less than or equal to 1000. The second example states that all the exhaust valves are either open or closed. The third states that all programs resident in main memory are suspended and not active.

Worked example 4.4 Convert the following extract of a system specification into predicate calculus.

> If the main processor is in quiescent state, then all processes are suspended and all critical devices will be down.

The solution uses two quantifiers.

Solution The predicate is

$$CompState(main, quiescent) \Rightarrow$$
$$\forall p : procs \bullet ProcState(p, suspended) \land$$
$$\forall d : CritDevs \bullet DevState(d, down)$$

■

Worked example 4.5 Translate the following predicate into natural language.

$$\forall p : ProgsInMemory \bullet ProgState(p, suspended) \Rightarrow$$
$$\forall pr : processors \bullet ProcessorState(pr, idle)$$

Solution The translation is

> If all the programs resident in memory are suspended, then all the processors are idle.

■

Exercise 4.5
Express the following predicates in natural language.

(i) $\forall d : ActiveDevices \bullet$
$CommStatus(d, transfer)$

(ii) $\forall monitor : AttachedMonitors \bullet$
$MonState(monitor, on) \Rightarrow TestState(monitor, DummyTest)$

(iii) $\forall monitor : AttachedMonitors \bullet MonState(monitor, on)$

(iv) $\forall device : ActiveDevices \bullet$
$DeviceState(device, on) \Rightarrow CompState(SateliteComputer, on)$

(v) $\forall d : ActiveDevices \bullet DeviceState(d, on) \Rightarrow$
$\forall p : DeviceProcesses \bullet ProcessState(p, running)$

(vi) $\forall file : Files \bullet$
$FileCategory(file, trans) \lor FileCategory(file, update)$

■

Exercise 4.6
Convert the following extracts from a system specification into predicate calculus.

(i) All the files in the file store will be active or passive but not both.
(ii) A reactor in the monitoring system will always be on-line if it is marked as functioning in the reactor table.
(iii) Each user has read or write access to the monitor file but not both.
(iv) If a file in the main store is marked for archiving, then it is also marked for deletion and for logging.
(v) Every process in every active queue is attached to the master table header.
(vi) A system contains three types of processes: active processes; suspended processes; and ready processes. If a process is active, then all the devices attached to that process are regarded as not available. If a process is suspended, then all the devices attached to that process are regarded as available. If a process is ready, no devices should be attached to the process.
(vii) If all the transactions have the first field blank, then the update file should be opened for overnight processing.
(viii) A computer system serves a number of users. Each user can be categorized as being a normal user or a super user but not both.
(ix) A system contains two types of processes: active processes and suspended processes. All suspended processes will be attached to the monitor queue.

∎

A statement involving universal quantification is, like a statement using existential quantification, a predicate. For example, if a monitor attached to a chemical reactor is malfunctioning and is generating a ready signal and it is asserted that

$$\forall\, mon : monitors \bullet \\ \quad MonitorState(mon, malfunctioning) \land \neg\, signal(mon, ready)$$

then it is obvious that the assertion is false.

Worked example 4.6 The following is an extract from a system specification. It is followed by a series of assertions written in predicate calculus. Indicate the truth or otherwise of each assertion.

> Users of a computer system are either normal users or privileged users. Files in the computer system can either be read-protected or write-protected. A normal user only has access to read-protected files, while the privileged user has access to both types of file.

$$\forall\, user : users \bullet \\ \quad UserType(user, normal) \land UserType(user, privileged)$$

$$\forall\, user : users;\ file : files \bullet$$
$$FileStatus(file, ReadProtected) \Rightarrow access(user, file)$$

$$\forall\, user : users;\ file : files \bullet$$
$$UserType(user, privileged) \Rightarrow access(user, file)$$

Solution The first assertion is false since it states that all users are both normal users and privileged users. The second assertion is true since it states that all users have access to all read-protected files. The third assertion is true as it states that all privileged users have access to all files. ■

Exercise 4.7
The following is an extract from a system specification. It is followed by five assertions written in predicate calculus. Indicate the truth or otherwise of the assertions.

There are three types of transactions which are to be processed by the system. They are: query transactions, update transactions, and verification transactions. Two sorts of programs are allowed to issue transactions. They are privileged programs and normal programs. Privileged programs are allowed to issue all types of transactions while normal programs are allowed to issue only query transactions. A program can be executed by any of the processors currently allocated to the system. However, if the processor is an Ia33 processor or an Ia22 processor, only update transactions are allowed.

(i) $\forall\, prog : programs \bullet$
$ProgStatus(prog, privileged) \lor ProgStatus(prog, normal)$

(ii) $\forall\, prog : programs \bullet$
$ProgStatus(prog, privileged) \Rightarrow issue(prog, update) \lor$
$issue(prog, verify) \lor issue(prog, query)$

(iii) $\forall\, prog : programs \bullet$
$(running(prog, Ia33) \lor running(prog, Ia22)) \land$
$(ProgStatus(prog, normal) \Rightarrow issue(prog, update))$

(iv) $\forall\, prog : programs \bullet$
$\neg\, running(prog, Ia33) \Rightarrow issue(prog, query)$

(v) $\forall\, prog : programs \bullet$
$\neg\, running(prog, Ia33) \land \neg\, running(prog, Ia22) \land$
$ProgStatus(prog, privileged) \Rightarrow issue(prog, update)$

■

There is a relationship between universal and existential quantification. Universal quantification expresses a property of *all* objects in a class while existential quantification asserts *at least one* object in a class has a property.

An assertion that not all objects in a class have a particular property is also an assertion that at least one object in the class does not have the property. Formally, this means that

$$\neg \forall \ldots \bullet \mathit{property} \Leftrightarrow \exists \ldots \bullet \neg \mathit{property}$$

Similarly, an assertion that there doesn't exist an object which has a property is equivalent to an assertion that all the objects do not have the property. Formally, this means that

$$\neg \exists \ldots \bullet \mathit{property} \Leftrightarrow \forall \ldots \bullet \neg \mathit{property}$$

Some examples of these equivalences follow. They are expressed as tautologies.

$$\neg \forall p : \mathit{processes} \bullet \mathit{ProcessState}(p, \mathit{active})$$
$$\Leftrightarrow$$
$$\exists p : \mathit{processes} \bullet \neg \mathit{ProcessState}(p, \mathit{active})$$

$$\neg \exists i : \mathbb{N} \bullet i^2 = 100 \Leftrightarrow \forall i : \mathbb{N} \bullet \neg i^2 = 100$$

$$\neg \exists rs : \mathit{reacts} \bullet \mathit{ReactorState}(rs, \mathit{active}) \wedge \mathit{LineStatus}(rs, \mathit{OnLine})$$
$$\Leftrightarrow$$
$$\forall rs : \mathit{reacts} \bullet \neg \mathit{ReactorState}(rs, \mathit{active}) \vee \neg \mathit{LineStatus}(rs, \mathit{OnLine})$$

The first example states that, if not all processes are active, it is equivalent to saying that at least one process is not active. The second example states that, if there isn't a natural number i whose square is equal to a hundred, this is equivalent to asserting that all the natural numbers do not have their square equal to a hundred. Although both these statements are patently false, they are still equivalent. The final example states that, if it is not true that there is a reactor which is both active and on-line, then it is equivalent to saying that all reactors are not on-line or not active. The statements are the same since

$$\neg (\mathit{ReactorState}(rs, \mathit{active}) \wedge (\mathit{LineStatus}(rs, \mathit{OnLine}))$$

is equivalent to

$$\neg \mathit{ReactorState}(rs, \mathit{active}) \vee \neg \mathit{LineStatus}(rs, \mathit{OnLine})$$

by the *DeM* $\wedge$ law. Intuition convinces us that the equivalences between universal and existential quantification hold. However, a proof is not difficult. The proof can be established by considering the meaning of both universal and existential quantification. When quantification is written as

$$\forall \ldots \bullet P$$

where P is a predicate, what is being asserted is that P is true for all combinations of the bound variables in the signature. Thus,

$$\forall \ldots \bullet P \Leftrightarrow P_1 \land P_2 \land P_3 \land \ldots \land P_n$$

where n is the number of different combinations of bound variables and P_i is the predicate formed by substituting each combination of bound variable. For example, if there are two reactors, *ReactorA* and *ReactorB*, and two monitors, *MonitorA* and *MonitorB*, then

$$\forall\, mon : monitors;\ react : reactors \bullet attached(mon, react)$$

is equivalent to

$$attached(ReactorA, MonitorA) \land attached(ReactorB, MonitorA)$$
$$\land\ attached(ReactorA, MonitorB) \land attached(ReactorB, MonitorB)$$

Similarly, existential quantification is equivalent to asserting a series of predicates separated by or operators

$$\exists \ldots \bullet P \Leftrightarrow P_1 \lor P_2 \lor P_3 \lor \ldots \lor P_n$$

where n is the number of different combinations of bound variables in the signature of the quantifier. For example, if there are three communication lines, *comm*1, *comm*2, and *comm*3, and two multiplexors, *mul*1 and *mul*2. The predicate

$$\exists\, comm : CommunicationLines \bullet LineState(comm, active)$$

is equivalent to

$$LineState(comm1, active) \lor LineState(comm2, active)$$
$$\lor\ LineState(comm3, active)$$

and

$$\exists\, comm : CommunicationLines;\ multi : multiplexors \bullet$$
$$\quad LineState(comm, active) \land connected(comm, multi)$$

is equivalent to

$$(LineState(comm1, active) \land connected(comm1, mul1)) \lor$$
$$(LineState(comm1, active) \land connected(comm1, mul2)) \lor$$
$$\lor \ldots \lor$$
$$(LineState(comm3, active) \land connected(comm3, mul2))$$

Given that

$$\forall \ldots \bullet P \Leftrightarrow P_1 \wedge P_2 \wedge P_3 \wedge \ldots \wedge P_n \cdot$$

and

$$\exists \ldots \bullet P \Leftrightarrow P_1 \vee P_2 \vee P_3 \vee \ldots \vee P_n$$

it is now possible to demonstrate their equivalence. If

$$\forall \ldots \bullet P \Leftrightarrow P_1 \wedge P_2 \wedge P_3 \wedge \ldots \wedge P_n$$

then

$$\neg (\forall \ldots \bullet P) \Leftrightarrow \neg (P_1 \wedge P_2 \wedge P_3 \wedge \ldots \wedge P_n)$$

Now

$$\neg (P_1 \wedge P_2 \wedge P_3 \wedge \ldots \wedge P_n)$$

is equivalent to

$$\neg P_1 \vee \neg P_2 \vee \neg P_3 \vee \ldots \vee \neg P_n$$

by the *DeM* $\wedge$ rule of inference, but

$$\neg P_1 \vee \neg P_2 \vee \neg P_3 \vee \ldots \vee \neg P_n$$

is equivalent to

$$\exists \ldots \bullet \neg P$$

thus,

$$\neg \forall \ldots \bullet P \Leftrightarrow \exists \ldots \bullet \neg P$$

Worked example 4.7 Demonstrate that

$$\neg \exists \ldots \bullet P \Leftrightarrow \forall \ldots \bullet \neg P$$

Solution First,

$$\exists \ldots \bullet P \Leftrightarrow P_1 \vee P_2 \vee P_3 \vee \ldots \vee P_n$$

The predicate

$$\neg \exists \ldots \bullet P$$

is equivalent to

$$\neg (P_1 \vee P_2 \vee P_3 \vee \ldots \vee P_n)$$

However,

$$\neg (P_1 \vee P_2 \vee P_3 \vee \ldots \vee P_n)$$

is equivalent to

$$\neg P_1 \wedge \neg P_2 \wedge \neg P_3 \wedge \ldots \wedge \neg P_n$$

by the *DeM* ∨ rule of inference. Now

$$\neg P_1 \wedge \neg P_2 \wedge \neg P_3 \wedge \ldots \wedge \neg P_n$$

is equivalent to

$$\forall \ldots \bullet \neg P$$

Hence

$$\neg \exists \ldots \bullet P \Leftrightarrow \forall \ldots \bullet \neg P$$

■

Until now universal and existential quantification have appeared singly in predicates. However, there is no reason why they cannot be mixed. For example, the predicate

$$\exists\, prog : programs \bullet \forall\, us : users \bullet HasAccess(us, prog)$$

states that there is at least one program to which all users have access. Predicates with both universal and existential quantifiers tend to be a little difficult to read. Exercise 4.8 gives you practice in understanding predicates. You would be well advised to complete it before proceeding further.

Exercise 4.8
Translate the following predicates into natural language.

(i) $\forall\, us : users \bullet \exists\, p : programs \bullet HasAccess(us, p)$
(ii) $\forall\, i : 1 \mathinner{..} 100 \bullet \exists\, j : 1 \mathinner{..} 100 \bullet i^2 = j$
(iii) $\exists\, m : monitors \bullet \forall\, l : CommsLines \bullet IsAttached(m, l)$
(iv) $\forall\, i : 1 \mathinner{..} 50 \bullet \exists\, j : 1 \mathinner{..} 10 \bullet i + j = 3$
(v) $\exists\, j : 1 \mathinner{..} 1000 \bullet \forall\, i : 1 \mathinner{..} 100 \bullet j = i$
(vi) $\exists\, file : FileStore \bullet \forall\, u : users;\ d : directories \bullet$
$\quad HasAccess(file, u) \wedge CanBeCopied(file, d)$
(vii) $\exists\, i : 1 \mathinner{..} 10 \bullet \forall\, j : 1 \mathinner{..} 10 \bullet \exists\, k : 1 \mathinner{..} 200 \bullet j^2 + i = k$
(viii) $\forall\, k : 1 \mathinner{..} 10 \bullet \exists\, p : 1 \mathinner{..} 5 \bullet \forall\, j : 1 \mathinner{..} 3 \bullet k * p > j$

■

4.4 Rules of inference and quantification

Chapter 3 has described how rules of inference can be used to derive proofs and simplify predicates. That chapter introduced a number of such rules of inference. You will not be surprised to know that there are a number of rules of inference which involve universal and existential quantifiers. A selection is shown below.

$$\frac{\forall x{:}S \bullet P(x), \forall x{:}S \bullet Q(x)}{\forall x{:}S \bullet P(x) \wedge Q(x)} \; = \text{ForAllAnd} \qquad \frac{\forall x{:}S \bullet P(x)}{\exists x{:}S \bullet P(x)} \; = \text{ForAllEx}$$

$$\frac{\forall x{:}S \bullet P(x)}{P(a)} \; = \text{ForAllOne} \qquad \frac{P(a)}{\exists x{:}S \bullet P(x)} \; = \text{OneEx}$$

$$\frac{\neg \; \forall x{:}S \bullet P(x)}{\exists x{:}S \bullet \neg \; p(x)} \; = \text{NotForAll} \qquad \frac{\neg \; \exists x{:}S \bullet P(x)}{\forall x{:}S \bullet \neg \; P(x)} \; = \text{NotEx}$$

A number of the rules assume that a is contained in the set S. All these rules can be used in the same way that the rules described in Chapter 3 can be used.

4.5 Predicate calculus and design specification

Enough mathematics has been described for a practical application to be described. Such an application is the specification of program units using predicate calculus. During system design a functional specification is transformed into a system design. Such a design will consists of discrete units. In turn these units will consist of modules which contain procedures and functions which are related by virtue of the tasks they perform. For example, in a system design for controlling a chemical plant there may be a module which is concerned with those processing functions associated with the thermocouples attached to the reactor. This module may contain procedures and functions which: read a temperature from a thermocouple; check on the functioning of a thermocouple; and sense and decode malfunctions which may arise in a thermocouple.

The normal way to specify the function of such procedures and functions is to use natural language. A typical example is

> The procedure *SumUpOut* has two parameters. The first parameter is an integer array *vals* which has a range $1, \ldots, 10$. The second parameter is a boolean variable *OutRange*. This is set true if any value in *vals* is less than or greater than 2000.

Given such a description a programmer can construct a program which satisfies the specification. The same drawbacks of using natural language in system specification are true for its use in design specification. Fortunately, predicate calculus, augmented by a few extra facilities, can be used as an exact notation for describing the function of program units.

The first extra facility is a predicate formed from the existential quantifier which asserts that only one object in a class holds. This quantifier is known as the **singular existential quantifier**, and it is written in the same form as the existential quantifier except for the fact that it is subscripted by 1. Thus,

$$\exists_1 x : \mathbb{N} \bullet x = 5$$

asserts that only one natural number equals 5. It is important to recognize that the singular existential quantifier is different from the existential quantifier that you are already familiar with. For example,

$$\exists_1 \, mon : monitors \bullet MonitorState(mon, off)$$

asserts that *exactly* one monitor is off, while

$$\exists \, mon : monitors \bullet MonitorState(mon, off)$$

asserts that *at least one* monitor is off. For example, if there were 100 monitors, there could be anything from one to 100 monitors that are in the off state.

The second extra facility is provided by a quantifier known as the **counting quantifier**. Again it is written in the same way as the existential and universal quantifiers. The counting quantifier is not a predicate. It does not deliver a true or false value but a natural number. This number represents the number of objects in a class which have a certain property. The quantifier is written as Ω. Some examples of its use are shown below.

$$\Omega x : 1 \mathinner{..} 20 \bullet x^2 < 17$$

$$\Omega mon : monitors \bullet \\ MonitorState(mon, off) \lor MonitorState(mon, functioning)$$

$$\Omega line : lines \bullet connected(line, reactor12)$$

$$\Omega x : 1 \mathinner{..} 100 \bullet \exists y : 1 \mathinner{..} 50 \bullet x + y < 10$$

The first example counts the number of values of x which lie between 1 and 20 whose squares are less than 17. There are in fact four $\{1, 2, 3, 4\}$. The second example gives the number of monitors which are either off or functioning. The third example gives the number of communication lines which are connected to *reactor*12. The final example gives the number of integers which lie between 1 and 100 which, when added to a number between 1 and 50, give a sum less than 10. The value of this will be eight since there are eight values of x which satisfy the predicate. They are $1, 2, 3, \ldots, 8$.

The final extra facility is the summation operator $\sum$. This returns the value of the sum of all objects in a class. The operator is written in the same way as the quantifiers previously described. Some examples of its use are

$$\sum i : 1 \mathinner{..} 10 \bullet i$$

$$\sum i : 1 \mathinner{..} 100 \bullet i^2$$

The first example gives the sum of the first 10 natural numbers while the second example gives the sum of the squares of the first 100 natural numbers.

Equipped with this slightly augmented form of the predicate calculus, it is relatively easy to construct precise specifications of the functions of program units. Such specifications will consist of two predicates. The first is called a **pre-condition**. It relates the values of global variables and parameters *before* a program is executed. The second predicate is called a **post-condition**. It relates the values of global variables and parameters *after* a program has been executed. As an example, consider the translation of the following specification into pre-condition and post-condition form.

> The function of the update procedure is to update the value of the global variable *SystemState*. The procedure has one integer parameter *temp*. If *temp* is greater than or equal to 200, then *SystemState* is set to zero; otherwise *SystemState* is set to a value of one. *SystemState* will have a value ranging between zero and five, *temp* will range from zero to 1000. *temp* will be unaffected by the procedure.

The predicate for the pre-condition can be written as

$$temp \geq 0 \land temp \leq 1000 \land SystemState \geq 0 \land SystemState \leq 5$$

The post-conditions should state that, if *temp* is greater than or equal to 200, then *SystemState* is set to zero; otherwise *SystemState* is set to one. This can be written as

$$\begin{aligned} & temp' = temp \\ & \land (temp \geq 200 \Rightarrow SystemState' = 0) \\ & \land (temp < 200 \Rightarrow SystemState' = 1) \end{aligned}$$

The primes over a variable indicate the value of the variable *after* a program unit has been executed. Thus, the post-condition above indicates that *temp* remains unchanged and that *SystemState* is given a value which is dependent on the value of *temp*.

Normally predicates such as $temp' = temp$ are omitted in post conditions. The fact that a variable such as $temp'$ is omitted in a predicate is sufficient to indicate that its value remains unchanged. However, if it is really necessary to make explicit that a variable value remains unchanged after execution of a program unit then such equalities will be used.

Worked example 4.8 A procedure *update* has two integer parameters. The function of the procedure is to examine its first parameter *flag* and set its second parameter *value*. If *flag* lies between zero and 10 then *value* is set to zero. However, if *flag* does not lie between zero and 10, then *value* is unchanged.

Write the corresponding post and pre-conditions.

Solution The question says nothing about the values of the parameters *flag* and *value* on entry to *update*. Therefore, no pre-condition can be constructed. The post-condition is

$$(flag \geq 0 \wedge flag \leq 10) \Rightarrow value' = 0$$

If the fact that *value* was unchanged if *flag* lay outside the range zero to 10 were specified then the predicate

$$(flag < 0 \vee flag > 10) \Rightarrow value' = value$$

would be conjoined to the above predicate

■

Exercise 4.9

Write pre- and post-conditions for the following design specifications.

(i) The function of the procedure *examine* is to examine its first parameter *val* and update its second integer parameter *NewVal*. If *val* is greater than zero, then it is added to *NewVal*, otherwise it is subtracted from *NewVal*. *val* will always have a value between -100 and 100.

(ii) The procedure *select* has three parameters. The first parameter *flag* is boolean; the second and third parameters *add* and *val* are integers. If the first parameter is true, then the second parameter is set to the modulus of the third parameter.

(iii) The procedure *QuotandRem* has four parameters: x, y, q, r. They are all natural numbers. The third parameter q will be set to the quotient of the first two parameters x and y, while the fourth parameter r will be set to the remainder upon dividing the first parameter by the second parameter. The first two parameters will be positive.

(iv) The procedure *NewVal* has one integer parameter *val*. The function of *NewVal* is to increment the parameter by one.

∎

Quantification can be used in both pre- and post-conditions. Indeed, if the properties of arrays need to be specified, then the use of quantification is the only way to achieve succinctness. For example, assume that a post-condition required that all the elements of an array $a[1..100]$ be zero. This would be written as

$$\forall i : 1..100 \bullet A[i] = 0$$

rather than

$$A[1] = 0 \wedge A[2] = 0 \wedge \ldots \wedge A[100] = 0$$

As an example of the use of quantifiers in a design specification, consider the following natural language description of the function of a procedure *addup*.

> The procedure *addup* has three parameters *arr*1, *arr*2, and *arr*3. All the parameters are integer arrays with range 1 .. 50. The function of *addup* is to add corresponding elements of *arr*1 and *arr*2 and place their sum in the corresponding element of *arr*3. On entry to *addup* all the elements of *arr*3 will be zero.

The pre-condition will be

$$\forall i : 1..50 \bullet arr3[i] = 0$$

the post-condition will be

$$\forall i : 1..50 \bullet arr3[i] = arr2[i] + arr1[i]$$

A more complicated example in the use of quantification occurs in connection with a sorting procedure *sort* whose natural language specification is

> The procedure *sort* has one parameter A. This is an integer array with a range 1 .. 40. The function of *sort* is to place the elements of A in ascending order. All the values contained in A will be less than 100 and greater than or equal to zero.

A first attempt at writing the pre-condition is

$$\forall i : 1 \mathinner{..} 40 \bullet A[i] \geq 0 \wedge A[i] < 100$$

while a first attempt at the post-condition is

$$\forall i : 1 \mathinner{..} 39 \bullet A'[i] \leq A'[i+1]$$

The pre-condition is correct but what about the post-condition? Certainly it states that the elements of A after *sort* has been executed will be in ascending order. However, there is an important component of the post-condition missing. This is the fact that the elements of A' after sorting match the elements of A, although they may not be in the same position in the array. One way of writing this might be

$$\forall i : 1 \mathinner{..} 40 \bullet \exists j : 1 \mathinner{..} 40 \bullet A'[j] = A[i]$$

This states that for every element in A there is an element in A' that matches it. The post-condition can then be written as

$$\begin{array}{l} \forall i : 1 \mathinner{..} 39 \bullet A'[i] \leq A'[i+1] \wedge \\ \forall i : 1 \mathinner{..} 40 \bullet \exists j : 1 \mathinner{..} 40 \bullet A'[j] = A[i] \end{array}$$

Is this correct? Unfortunately not. If A just contained ones and A' had a one in its first element and two in the rest of its elements the post-condition would hold since for each of the elements in A there exists an element in A' that matches it, but A' does not represent the sorted version of A. If A contained distinct elements, i.e. there was no duplication, then the post-condition above would be adequate. However, the pre-condition does not state this. This is an excellent example of the use of mathematics to clarify and question a specification.

An adequate post-condition would have to include a predicate which asserts that A' is a permutation of A, i.e. the number of occurrences of an integer in A matches the number of occurrences in A'. A predicate that asserts this is

$$\forall j : 0 \mathinner{..} 99 \bullet ((\Omega i : 1 \mathinner{..} 40 \bullet A'[i] = j) = (\Omega k : 1 \mathinner{..} 40 \bullet A[k] = j))$$

This gives the correct post-condition as

$$\begin{array}{l} \forall i : 1 \mathinner{..} 39 \bullet A'[i] \leq A'[i+1] \wedge \\ \forall j : 0 \mathinner{..} 99 \bullet ((\Omega i : 1 \mathinner{..} 40 \bullet A'[i] = j) = (\Omega k : 1 \mathinner{..} 40 \bullet A[k] = j)) \end{array}$$

Worked example 4.9 If the pre-condition for the sorting example presented above had to indicate that, at most, A contained only one occurrence of an integer, how would you write the pre-condition?

Solution The pre-condition would still have to state that all the elements of A ranged between zero and 100. However, this would have to be augmented with the fact that A contained only one occurrence of an integer. Since the range of A is $1..40$ and the integers lie between 1 and 100, some integers will not occur at all. The pre-condition will hence be

$$\forall\, i : 1..40 \bullet A[i] < 100 \wedge A[i] \geq 0 \wedge$$
$$\forall\, k : 0..99 \bullet (\Omega i : 1..40 \bullet A[i] = k) \leq 1$$

■

Worked example 4.10 A procedure *search* has three parameters: *vals*, *searchfor*, and *found*. *vals* is an integer array with a range of $1..200$. *searchfor* is an integer parameter. If *searchfor* occurs in *vals* then *found* is set to one; otherwise it is set to two. Write down the pre- and post-conditions

Solution Unfortunately, the natural language specification does not say anything about the values that the parameters have on entry to search. Hence a pre-condition cannot be written. The post-condition is

$$(\exists\, i : 1..200 \bullet vals[i] = searchfor \Rightarrow found' = 1) \wedge$$
$$(\neg\, \exists\, i : 1..200 \bullet vals[i] = searchfor \Rightarrow found' = 2)$$

■

Exercise 4.10

Write down the pre- and post-conditions for the following design specifications expressed in natural language.

(i) The procedure *examine* has four parameters. The first parameter is an integer array *vals* which has a range $1..20$. The other parameters: *SearchforRange*, *obj*, and *flag* are integers. *examine* looks at the first *SearchforRange* elements of *vals* and sets *flag* to one if at least one element is equal to *obj*.

(ii) The procedure *descend* has two parameters. The first is *valarr* which is an integer array with range $1..50$. The second is an integer *FlagSet*. If *valarr* is in descending order then *FlagSet* will be set to zero; otherwise it remains unchanged.

(iii) The procedure *clear* has three parameters. The first parameter *exam* is an integer array with a range $1..50$. The second is an integer array *ToBeCleared* with a range of 1 .. 75. The third parameter is an integer *obj* which lies between 0 and 200. If *obj* is contained in *exam*, then all the elements of *ToBeCleared* are set to zero; otherwise it remains unchanged.

(iv) The procedure *SameTime* has three parameters: *op*1, *op*2, and *clear*. These are all integer arrays with a range of 1 .. 100. *op*1 and *op*2 all contain positive integers less than or equal to 500. The function of *SameTime* is to examine *op*1 and *op*2 and, if they are identical, to set all the elements of *clear* to zero; otherwise it remains the same.

(v) The procedure *plateau* has two parameters. The first parameter *vals* is an integer array which has a range 1 .. 500. The second parameter *flag* is an integer. If *vals* contains a sequence of equal integers the length of which is greater than one, then *flag* is set to one; otherwise *flag* stays unchanged. *vals* contains positive integers less than or equal to 100.

■

5

Set theory

Aims

- To introduce the idea of a set as a collection of objects.
- To describe the important set operators.
- To outline how sets and predicate calculus can be combined to produce mathematical specifications.
- To describe some rules of inference involving sets.

5.1 Introduction

The previous sections of this book have concentrated on using propositional or predicate calculus for describing the functions of a software system. For example, you have now been given the mathematical tools to express the essence of the paragraph

> If all the monitoring valves are functioning and at least one computer is on-line, then, provided the system is in a functioning state, the header line will be displayed on all the VDUs attached to the system.

However, the mathematical tool-kit is not yet complete. Consider the paragraphs

> There are two types of file: user files and system files. A file cannot be both a user file and a systems file.
>
> All the files which are user files cannot be deleted.
>
> If a file is an archive file, then it cannot be either a user or system file.

It is possible to express these paragraphs using predicates such as *owner*. For example, the first paragraph can be expressed in predicate calculus as

$$\forall file : files \bullet \neg (FileType(file, user) \land FileType(file, system))$$

However, the concept of a class turns up time and time again in specifications. Consequently, it is worth introducing symbols, operators, and laws to cater for this. This is the role of **set theory**.

5.2 Sets and subsets

A set is a collection of objects. The objects may be natural numbers, names of files, locations of monitoring instruments, user names, or whatever objects are of interest to the specifier. The sets and the objects in a set are of a certain type in the sense that variables in a programming language such as Pascal are of a certain type. As will be seen later, this means that, as in modern programming languages, certain operators are valid only when their operands are of the same type.

A set can be specified as a collection of objects surrounded by curly brackets. Thus

$$\{21, 7, 14, 3\}$$

is an example of a set of objects which are natural numbers and

$$\{archiver, sorter, editor, finder\}$$

is a set of utility programs. The important property of a set is that **duplicates are not allowed**. Thus

$$\{1, 3, 4, 1, 9, 3\}$$

is not an example of a set.

A set can contain not only single-element objects but can also contain aggregates of objects. For example,

$$\{(1, 2), (3, 4), (4, 5)\}$$

is a set of pairs of natural numbers and

$$\{(mon1, mon2, mon3), (mon2, mon4, mon5), (mon3, mon7, mon8)\}$$

is a set of triples which contain monitor names.

A set can be **finite** or **infinite**.

$$\{1, 3, 5\}$$

is an example of a finite set, while the set of all the natural numbers is an example of an infinite set.

The objects which make up a set are known as its **members**. The fact that an object is a member of a set is written as

$$x \in S$$

If x does not belong in a set, then this is written as

$$x \notin S$$

Thus,

$$3 \in \{3, 4, 7\}$$
$$5 \in \{17, 230, 46, 5\}$$
$$update \in \{update, write, read\}$$

are examples of predicates which are true and

$$1 \notin \{2, 5, 7\}$$
$$vdu1 \notin \{vdu3, vdu8, vdu9\}$$

are also examples of predicates which are true.

Worked example 5.1 The following is an example of an extract from a natural language specification.

> If the file is a read-access file or not an update-access file then the system manager will regard the file as a non-archivable file.

Convert it into predicate calculus using the $\in$ and $\notin$ operators.

Solution

$$file \in ReadAccessFiles \vee file \notin UpdateAccessFiles \Rightarrow$$
$$file \in NonArchivableFiles$$

This illustrates the fact that set operators can be used in predicates and also that an object can be a member of a number of sets.

■

Exercise 5.1

Convert the following paragraphs of natural language specification into predicate calculus. Use the operators $\in$ and $\notin$ where necessary.

(i) All the files in the system will be read-access files or write-access files.
(ii) No file in the system will be an active file and a passive file, provided

the monitoring system is in normal condition status.
(iii) There is only one file which is a modify-file and an update-file.
(iv) There will be five system files in the system.
(v) The password2 file will contain the passwords of the users who are system users.

∎

5.2.1 Set specification

How is it possible to specify members of a set? One method has already been described where the elements of a set were explicitly listed. For example,

$$\{1, 3, 5\}$$

is an explicit listing of the set containing those odd numbers less than 6. Unfortunately, listing a set this way has a number of disadvantages. The first disadvantage is that, for large finite sets, explicit listing is tedious and, for infinite sets, impossible. The second disadvantage is that such a listing does not make a relationship between the elements of a set clear. For example, does the set

$$\{1, 3, 5\}$$

represent the odd numbers less than six or the positive square root of the elements of the set

$$\{1, 9, 25\}?$$

In order to cope with these disadvantages mathematicians have developed a notation which enables a set to be succinctly and unambiguously defined. An example of this is

$$\{\ n : \mathbb{N} \mid n^2 < 25 \bullet n\ \}$$

It defines a set of natural numbers whose squares are less than 25, i.e. it specifies the set

$$\{0, 1, 2, 3, 4\}$$

This way of defining a set is known as a **comprehensive specification**. Its general form is

$$\{\ \mathsf{Signature} \mid \mathsf{Predicate} \bullet \mathsf{Term}\ \}$$

Thus, the comprehensive specification

$$\{ \, n : \mathbb{N} \mid n < 20 \land n > 10 \bullet n \, \}$$

denotes the set of natural numbers which lie between 10 and 20, $n : \mathbb{N}$ is the signature, $n < 20 \land n > 10$ is the predicate, and n is the term.

A signature consists of a series of identifiers together with the sets to which they belong. Thus,

$$n : \mathbb{N}$$

states that n is a natural number and

$$a, b, c : \mathbb{N}$$

states that a, b, and c are all natural numbers. There is no restriction on the number and type of each identifier; for example,

$$x, y, z : \mathbb{N};\ \mathit{file}1 : \mathit{SystemFiles};\ \mathit{file}2 : \mathit{UserFiles}$$

is a signature which introduces three identifiers x, y, and z which belong to the set of natural numbers, *file*1 which belongs to the set *SystemFiles*, and *file*2 which belongs to the set *UserFiles*.

The predicate part of a comprehensive set specification defines the properties of the members of the set which is specified. Thus,

$$\{ \, n : \mathbb{N} \mid n^3 > 10 \bullet n \, \}$$

specifies the set of natural numbers which have the property that their cubes are greater than 10.

The term part of a comprehensive specification defines the form of the members of the set. A term consists of an expression which, when evaluated, will deliver a value which is of the same type as the set. For example,

$$\{ \, n : \mathbb{N} \mid n > 20 \land n < 100 \bullet n \, \}$$

states that a set will contain single natural numbers which satisfy the predicate $n > 20 \land n < 100$, while

$$\{ \, x, y : \mathbb{N} \mid x + y = 100 \bullet (x, y) \, \}$$

specifies the set of pairs which are natural numbers whose sum is 100, i.e.

$$\{(0, 100), (1, 99), (2, 98), \ldots, (100, 0)\}$$

Thus, the term in this example defines the fact that elements of the set are pairs. Some more examples of comprehensive set specifications with their natural language equivalents are now given.

$$\{ x : \mathbb{N} \mid x = 3 \bullet x \}$$

is the set of all natural numbers whose elements are equal to 3. This specifies the set $\{3\}$.

$$\{ x, y : \mathbb{N} \mid x + y = 5 \bullet x^2 + y^2 \}$$

is the set of natural numbers of the form $x^2 + y^2$ where $x + y$ is equal to 5. It represents the set

$$\{25, 17, 13\}$$

The set

$$\{ mon : monitors \mid MonitorState(mon, on) \bullet mon \}$$

specifies the set of monitors which are on.

$$\{ file : SystemFiles \mid file \in DeletedFiles \wedge file \in ArchivedFiles \bullet file \}$$

specifies the set of system files which are both deleted files and archived files.

If a term involves simple elements such as a, (a, b), or (a, b, c), then the term is normally omitted since it can be deduced from the predicate. For example, the set specification

$$\{ n : \mathbb{N} \mid n < 10 \wedge n > 5 \bullet n \}$$

would normally be written as

$$\{ n : \mathbb{N} \mid n < 10 \wedge n > 5 \}$$

and the set specification

$$\{ a, b : \mathbb{N} \mid a < 10 \wedge b < a \bullet (a, b) \}$$

would normally be written as

$$\{ a, b : \mathbb{N} \mid a < 10 \wedge b < a \}$$

with the term omitted.

Worked example 5.2 The set *oddten* is

$$\{1, 3, 5, 7, 9\}$$

Write down the comprehensive set specification of a set which is formed by extracting those members of *oddten* which are in the range 3 to 7.

Solution The set that is asked for is that whose members (x) are in *oddten* and which satisfy the predicate $x \geq 3 \wedge x \leq 7$. The full specification is

$$\{ x : oddten \mid x \geq 3 \wedge x \leq 7 \}$$

Again the term is omitted.
■

Worked example 5.3 Write down the comprehensive specification of the set of pairs for which the first element is a natural number greater than 10 and the second element is the square of the first element.

Solution The specification is

$$\{ n : \mathbb{N} \mid n > 10 \bullet (n, n^2) \}$$

Again there is a reason for writing the term.
■

Exercise 5.2
Define the following sets using a comprehensive specification omitting the term if necessary.

(i) The set of natural numbers less than 10.
(ii) The set of pairs of natural numbers whose sum of squares is less than 200.
(iii) The set of natural numbers which are between 10 and 53.
(iv) The set of pairs of the form (x, x^2) where x is greater than 100.
(v) The set of files which are both user files and system files.
(vi) The set of files which are user files and owned by Thomas.
(vii) The set of files which are user files and are accessed by more than 10 users.
(viii) The set of files which are not user files and not system files and which are in the directory *ControlX*.

■

Exercise 5.3
If *SystemFiles* is the set

$$\{Scheduler, FileHandler, CommandHandler, PeriphHandler\}$$

and *MasterFiles* is the set

$$\{Scheduler, FileHandler, UsFile1, UsFile2, UsFile3\}$$

which of the following predicates are true and which are false?

(i) $Scheduler \in SystemFiles$
(ii) $Scheduler \in SystemFiles \wedge Scheduler \notin MasterFiles$
(iii) $Scheduler \in \{ file : files \mid file \in SystemFiles \wedge file \notin MasterFiles \bullet file \}$
(iv) $Scheduler \in \{ file : files \mid file \in SystemFiles \wedge file \in MasterFiles \bullet file \}$
(v) $PeriphHandler \in \{ file : files \mid file \in SystemFiles \wedge file \in MasterFiles \}$

∎

5.2.2 The empty set

Consider the comprehensive specification

$$\{ n : \mathbb{N} \mid n > 10 \wedge n < 3 \}$$

This is the set of natural numbers which are greater than 10 and less than 3. How many elements would this set contain? Unfortunately, there is no natural number which satisfies the predicate of the specification. Thus there are no elements in this set. This set is known as the **empty set**. In this book it will be written as $\varnothing$. In some mathematics books it is written as $\{ \}$.

5.2.3 Subsets and the power set

Within any given set A there exists other sets which can be obtained by removing some of the elements of A. These are called **subsets** of A. For example, if A is

$$\{1, 3, 9, 14, 200\}$$

then both $\{1, 3, 9\}$ and $\{1, 9, 14, 200\}$ are subsets of A. The fact that a set is a subset of another set is expressed by the operators: $\subset$ and $\subseteq$. The predicate

$$A \subseteq B$$

is true if A is a subset of B including being equal to B. Thus,

$$\{1, 2\} \subseteq \{1, 2, 3, 4\}$$
$$\{1\} \subseteq \{1, 2, 3, 4\}$$
$$\{1, 2, 3, 4\} \subseteq \{1, 2, 3, 4\}$$

are all true. The predicate

$$A \subset B$$

is true when A is a **proper subset** of B, that is, A is a subset of B but not equal to B. Thus, the predicates

$$\{1,2\} \subset \{1,2,3,4\}$$
$$\{4\} \subset \{1,2,3,4\}$$
$$\{1,4\} \subset \{1,2,3,4\}$$

are all true, while

$$\{1,2,3,4\} \subset \{1,2,3,4\}$$

is false.

Worked example 5.4 What is the value of the predicate

$$\{4\} \subset \{\ n : \mathbb{N} \mid n > 3 \bullet n^2\ \}?$$

Solution The value is false since the comprehensive set specification describes the squares of the natural numbers greater than 3.

■

One curious result involving subsets and the empty set which confuses many students who start studying pure mathematics is that the empty set is a subset of *every* set.

Exercise 5.4

Which of the following predicates are true and which are false?

(i) $\{file1, file2\} \subset \{file1, file2, file3\}$
(ii) $\{\ n : \mathbb{N} \mid n < 4\ \} \subset \{\ n : \mathbb{N} \mid n > 7\ \}$
(iii) $\varnothing \subset \{\ file : files \mid FileState(file, on) \wedge FileState(file, off)\ \}$
(iv) $\{\ a, b : \mathbb{N} \mid a + b = 4 \bullet a^2 + b^2\ \} = \varnothing$
(v) $\{\ n : \mathbb{N} \mid n > 10\ \} \subset \{\ n : \mathbb{N} \mid n > 10\ \}$
(vi) $\{\ n : \mathbb{N} \mid n^2 \leq 100\ \} \subset \{\ n : \mathbb{N} \mid n \leq 10\ \}$
(vii) $\{3,4\} \subset \{7,8,9\} \Rightarrow \{2,4\} \subset \{2,3,4\}$
(viii) $\{3\} \subset \{\ n : \mathbb{N} \mid n > 2 \bullet n\ \} \vee \{3\} \subset \{4,5,6\}$
(ix) $\{file1, file2\} \subset \{file2, file3\} \wedge \{file1, file2\} \subset \{file1, file3\}$

■

The set of all possible subsets of a set A is known as the **power set** of A.

The concept of a power set occurs time and time again in mathematics and in system specifications. Because of that it is often given a special symbol $\mathbb{P}$. The power set of A is formally written $\mathbb{P}\ A$. The power set of any set A is defined formally by means of the predicate

$$a \in \mathbb{P}\ A \equiv a \subseteq A$$

Thus, $\mathbb{P}\{1,2,3\}$ is

$$\{\varnothing, \{1\}, \{2\}, \{3\}, \{1,2\}, \{1,3\}, \{2,3\}, \{1,2,3\}\}$$

remembering, of course, that the empty set is a subset of *any* set.

Worked example 5.5 Write down $\mathbb{P}\{\mathit{file}1, \mathit{file}2\}$

Solution Since the power set of a set is the set of all combinations of the members of that set plus the empty set, the answer is

$$\{\varnothing, \{\mathit{file}1\}, \{\mathit{file}2\}, \{\mathit{file}1, \mathit{file}2\}\}$$

Don't forget the empty set.
■

Worked example 5.6 What is the value of the predicate $\{\ n : \mathbb{N} \mid n > 3\ \} \subset \mathbb{P}\ \mathbb{N}$?

Solution The value is true since any set of natural numbers will be included in the powerset of the natural numbers.
■

5.3 Set operators

The previous section introduced the concept of a set as a collection of distinct objects or elements which have a certain type. To manipulate such sets requires a series of operators. Already, the operators $\subset$, $\subseteq$, $\in$, $\notin$, and $\mathbb{P}$ have all been described. The operator $\in$ can be regarded as a primitive as all other set operators can be defined in terms of it using predicate calculus.

5.3.1 Set equality

A new operator which has been used informally in the previous section is the equality operator. This is written as $=$. Its operands must be of the same type. If A and B are two sets of the same type, then $A = B$ is true when each set contains the same elements. Thus,

$$\{1,2,4,5\} = \{1,2,5,4\}$$
$$\{9,8,7,6,5\} = \{7,8,6,9,5\}$$

are true while

$$\{1,3,5\} = \{1,3,7,11\}$$
$$\{1,3,5\} = \{1,3,5,9,11\}$$

are false. Again $=$ can be defined in terms of $\in$ and predicate calculus:

$$A = B \equiv \forall\, a : A \bullet a \in B \wedge \forall\, b : B \bullet b \in A$$

Given this definition of $=$, it is possible to define the $\subset$ operator as

$$A \subset B \equiv \forall\, a : A \bullet a \in B \wedge \neg\,(A = B)$$

which just states that A is a proper subset of B when every element of A occurs in B and A is not equal to B.

5.3.2 Set union and set intersection

System specifications often contain sentences such as

> Those monitors which are in the on state and are also attached to a critical reactor will be defined in this document as critical monitors.
>
> The collection of files which are in the system directory and which are accessible by users with a level 2 status are known as low-level files.
>
> Communication lines which are quiescent or which are active will be referred to as current communication lines.

Such sentences define new sets by combining previously defined sets. There is hence a need for set operators which form new sets from existing sets. The two most important operators which perform this function are $\cup$ and $\cap$. $\cup$ is the **union operator**. Both its two arguments are sets. It forms the set whose elements are in *either* of its arguments. Thus,

$$\{1,2,3\} \cup \{3,4,5,6\}$$

is

$$\{1,2,3,4,5,6\}$$

$\cap$ is the **intersection operator**. Again this operator has two arguments. It forms the set whose elements occur in *both* of its arguments. Thus,

$$\{3,4,7,9,11\} \cap \{7,11,13,15\}$$

is

$$\{7,11\}$$

Both the union and intersection operators take as operands sets which are of the same type, i.e. they contain objects of the same type. Thus, it is not legal to write

$$\{1, 2, 3\} \cap \{monitor1, monitor2\}$$

The union operator can be formally defined using $\in$ and predicate calculus as

$$A \cup B = \{\ x : T \mid x \in A \lor x \in B\ \}$$

where T is the type of the objects which make up the set, for example, the set $\mathbb{N}$ or the set of all functioning reactors.

Worked example 5.7 Formally define the $\cap$ operator.

Solution The intersection operator forms a set whose elements are in both of its arguments

$$A \cap B = \{\ x : T \mid x \in A \land x \in B\ \}$$

This is achieved by the use of the $\land$ operator in the set specification.
■

Exercise 5.5
Which of the following predicates are true and which are false?

(i) $\{3, 4, 5\} \cup \varnothing = \varnothing$
(ii) $\{3, 4, 5\} \cap \varnothing = \varnothing$
(iii) $\{\ n : \mathbb{N} \mid n^2 < 20\ \} \cap \{1, 2, 3\} = \{2, 3\}$
(iv) $\{\ n : \mathbb{N} \mid n > 5 \land n < 10\ \} \cup \{\ n : \mathbb{N} \mid n \leq 5\ \} = \{\ n : \mathbb{N} \mid n < 10\ \}$
(v) $\{mon1\} \subset \{mon2, mon3\} \cup \{mon1, mon2, mon4\}$
(vi) $\{mon1\} \subset \{mon2, mon3\} \cap \{mon1, mon2, mon3\}$
(vii) $\{\ mon : monitors \mid on(mon)\ \} \cap \{\ mon : monitors \mid off(mon)\ \} =$
$\{\ mon : monitors \mid on(mon) \land off(mon)\ \}$
(viii) $\{\ line : CLines \mid high(line)\ \} \subset \{\ line : CLines \mid high(line) \lor off(line)\ \}$

■

5.3.3 Set difference

Often in a system specification there will be sentences which describe the fact that a set of objects are like another set of objects but not like another set. For example,

Global files are those system files which are not secure files.

Working lines are those communications lines which are not quiescent.

The set operator which is used to model these statements is the **set difference operator** $\setminus$. It has two arguments both of the same type. It forms the set which is its first argument with elements of the second argument removed. Thus,

$$\{1,2,4,8,9\} \setminus \{1,2,3\} = \{4,8,9\}$$
$$\{1,2,3\} \setminus \{1,2,3\} = \varnothing$$

5.3.4 The cross product

The cross product of two sets A and B is denoted by

$$A \times B$$

The operator forms the set of pairs where the first element of each pair is drawn from A and the second element is drawn from B. Thus,

$$\{1,2,3\} \times \{4,5\} = \{(1,4),(1,5),(2,4),(2,5),(3,4),(3,5)\}$$

and

$$\{line1, line2\} \times \{10,12,3\} =$$
$$\{(line1,10),(line1,12),(line1,3),(line2,10),(line2,12),(line2,3)\}$$

Formally, the cross product is defined as

$$A \times B = \{\ a : A;\ b : B \mid (a, b)\ \}$$

The notion of a cross product can be generalized for any number of operands

$$A_1 \times A_2 \times A_3 \times \ldots \times A_n =$$
$$\{\ a_1 : A_1;\ a_2 : A_2;\ a_3 : A_3;\ \ldots a_n : A_n \mid (a_1, a_2, a_3 \ldots a_n)\ \}$$

Thus,

$$\{1,2\} \times \{3,4\} \times \{line1, line2\}$$

is

$$\{(1,3,line1),(1,3,line2),(1,4,line1),(1,4,line2),$$
$$(2,3,line1),(2,3,line2),(2,4,line1),(2,4,line2)\}$$

5.3.5 Set cardinality

The **cardinality** of a set is the number of elements in the set. For example, the cardinality of the set

$$\{tax, update, oldupdate\}$$

is 3. In this book the operator # is used as the cardinality operator. When applied to a set it gives the number of elements in the set. For example,

$$\#\{old, new, medium, fast, slow\} = 5$$

and

$$\#\{\ n : \mathbb{N} \mid n < 4\ \} = 4$$

are both true predicates.

5.4 Reasoning and proof in set theory

In the same way that it is possible to reason and derive theorems in predicate calculus, it is possible to reason and derive theorems in set theory using rules of inference. Some rules of inference are now shown.

A, B, and C are sets of the same type while x is an object of the same type as the objects contained in A, B, and C.

$\frac{x \in A, x \in B}{x \in A \cap B}$ = *in* ∩ $\qquad$ $\frac{x \in A}{x \in A \cup B}$ *in* ∪

$\frac{x \in A, A \subset B}{\{x\} \subset B}$ *in* ⊂ $\qquad$ $\frac{A \subset B, B \subset C}{A \subset C}$ *tran* ⊂

$\frac{A = B, B = C}{A = C}$ *tran* =

A number of theorems involving sets can also be stated. A selection is shown below.

$\vdash A \cup A = A$
$\vdash A \cup \varnothing = A$
$\vdash A \cup B = B \cup A$
$\vdash (A \cup B) \cup C = A \cup (B \cup C)$
$\vdash A \cap \varnothing = \varnothing$
$\vdash A \cap B = B \cap A$
$\vdash A \cap A = A$
$\vdash (A \cap B) \cap C = A \cap (B \cap C)$
$\vdash A \cup (B \cap C) = (A \cup B) \cap (A \cup C)$
$\vdash A \cap (B \cup C) = (A \cap B) \cup (A \cap C)$

These theorems can all be proved by recourse to the formal definition of operators such as ∪ and ∩.

5.5 Modelling a file system

To conclude this chapter an example of using set theory and predicate calculus for system specification is presented. It shows that a relatively realistic example can be modelled with the limited mathematical tool-kit which this book has so far presented. Each paragraph of the system specification is written in natural language; it is followed by the mathematical specification in terms of sets, set operators and predicates. The prime notation introduced in the previous chapter is used to describe the effect of events on the mathematical structures used to model the system. The whole specification is based on the following informal statement of requirements.

> The purpose of the filing system is to keep and maintain a series of files belonging to a number of users of a computer system. There will be two different types of user: normal users and system users. The latter will have extra privileges such as the ability to access user files and introduce new users into the system.

The specification is written as a series of paragraphs.

> 1. There will be two different types of users: normal users and system users. The latter will have extra privileges.

Assume first that there is a set of all possible user names *AllUserNames*. Also, there will be a set *users* which will contain all the user names of current users of the system. Furthermore, there will be two sets of names *NormalUsers* and *SystemUsers*. These will hold each category of user. Assuming that there are no other type of user and that a user cannot be simultaneously a normal user and a system user, then this paragraph can be written as

$$
\begin{aligned}
&users \subseteq AllUserNames \land \\
&NormalUsers \cup SystemUsers = users \land \\
&NormalUsers \cap SystemUsers = \varnothing
\end{aligned}
$$

The next paragraph restricts the number of files that a user can own.

> 2. Each user is allowed to own a number of files. These are stored in the file store of the computer. Each normal user can own up to 50 files while a system user can own up to 100 files.

This paragraph opens up a number of questions concerning how to model the file store of the computer. One possible solution is to represent it as a set of file names. Unfortunately, there are two major drawbacks to this. First, there is a probability that two users will give the same name to two different files. Since a set only contains distinct elements, there would be no way of modelling these files as sets. Second, there is no way to relate each file in the file store to the owner of the file.

A better solution is to model the file store as a set of pairs. Each pair consists of the name of a user and the name of the file owned by the user. If *AllFileNames* is the set of all possible valid file names, then the file store can be written as

$$FileStore \in \mathbb{P}\,(AllUserNames \times AllFileNames)$$

The restriction about the number of files which each user can own can then be written as

$$\begin{array}{l} \forall\, us : NormalUsers \bullet \\ \quad \#\{\, file : AllFileNames \mid (us, file) \in FileStore \,\} \leq 50 \wedge \\ \forall\, us : SystemUsers \bullet \\ \quad \#\{\, file : AllFileNames \mid (us, file) \in FileStore \,\} \leq 100 \end{array}$$

The next paragraph details the status of files.

> 3. Each file can be in two states: public or private. A public file can be read by all users and a private file can only be read by the user that owns it. A file which a user wants to be stored on magnetic tape is known as an archive file

Assuming that a file cannot simultaneously be a private file and a public file, then this paragraph can be written as

$$\begin{array}{l} ArchiveFiles \subseteq FileStore \wedge \\ PrivateFiles \cap PublicFiles = \varnothing \wedge \\ PrivateFiles \cup PublicFiles = FileStore \end{array}$$

> 4. Users of the filing system can perform any of the following operations
>
> | | |
> |---|---|
> | CREATE | Creates a new file, |
> | DELETE | Deletes an existing file, |
> | SET-PUBLIC | Places a file in a public state, |
> | SET-PRIVATE | Places a file in a private state, |
> | ARCHIVE | Marks a file for archiving, |
> | DE-ARCHIVE | Reverses the effect of the ARCHIVE command, |
> | CREATE-USER | Introduces a new user into the system. |
>
> Each command will normally refer to a file or a user.

During the operation of the file store a number of events will occur which leave all the elements of the file store unchanged. For example, a user mistyping a command will not affect the file store. In order to describe this a predicate *unchanged* is used which is defined as

$$SystemUsers' = SystemUsers \land NormalUsers' = NormalUsers \land$$
$$FileStore' = FileStore \land PublicFiles' = PublicFiles \land$$
$$PrivateFiles' = PrivateFiles \land ArchiveFiles' = ArchiveFiles$$

5. The CREATE command creates a new file in the file store. This command can be used both by normal users *and* system users. The effect of this command is to add a file to the file store. This file will be owned by the user who issued the command. The file so created will have private status.

In order to specify the effect of this command it will be necessary to denote the user who types in the command as *user* and the file to be created as *FileName*. The effect of the CREATE command is

$$\begin{aligned}
&command = CREATE \Rightarrow \\
&\quad (FileStore' = FileStore \cup \{(user, FileName)\} \land \\
&\quad PrivateFiles' = PrivateFiles \cup \{(user, FileName)\} \land \\
&\quad NormalUsers' = NormalUsers \land \\
&\quad SystemUsers' = SystemUsers \land \\
&\quad PublicFiles' = PublicFiles \land \\
&\quad ArchiveFiles' = ArchiveFiles)
\end{aligned}$$

There are two important points to notice about this predicate. First, primes are used to indicate the value of an object *after* a command has been executed. In the case of the CREATE command above, the sets *PrivateFiles* and *FileStore* are modified by the CREATE command and hence appear with primes. Second, the sets *NormalUsers*, *SystemUsers*, *PublicFiles*, and *ArchiveFiles* are unaffected by the correct operation of the CREATE command.

6. The DELETE command removes an existing file from the file store. The command can be used by both normal and system users. If a system user types the command, then any file can be deleted from the file store. If a normal user types the command, then only a file owned by the user can be deleted from the file store

There is a slight complication when this command is typed in by a system user. In order to identify uniquely a file not only must the file name be provided but also the name of the user whose file is to be deleted. No clue can be gained about the way that this name is communicated to the filing system. The specification shall assume that a user name *usname* is provided by the system user. The effect of the DELETE command is

$$\begin{aligned}
&command = DELETE \Rightarrow\\
&\quad (user \in SystemUsers \Rightarrow\\
&\qquad (FileStore' = FileStore \setminus \{(usname, FileName)\}) \wedge\\
&\quad user \in NormalUsers \Rightarrow\\
&\qquad (FileStore' = FileStore \setminus \{(user, FileName)\}) \wedge\\
&\quad NormalUsers' = NormalUsers \wedge\\
&\quad SystemUsers' = SystemUsers \wedge\\
&\quad PrivateFiles' = PrivateFiles \wedge\\
&\quad PublicFiles' = PublicFiles \wedge\\
&\quad ArchiveFiles' = ArchiveFiles)
\end{aligned}$$

The next paragraph defines the SET-PUBLIC command.

> 7. The SET-PUBLIC command sets the state of an existing file to be public. The command can be used by both normal and system users. If a system user types the command, then any file can be marked public. If a normal user types the command, then only a file owned by that user can be made public.

The effect of the SET-PUBLIC command is

$$\begin{aligned}
&command = SET - PUBLIC \Rightarrow\\
&\quad (user \in NormalUsers \Rightarrow\\
&\qquad (PrivateFiles' = PrivateFiles \setminus \{(user, Filename)\} \wedge\\
&\qquad PublicFiles' = PublicFiles \cup \{(user, FileName)\})\\
&\quad \wedge\ user \in SystemUsers \Rightarrow\\
&\qquad (PrivateFiles' = PrivateFiles \setminus \{(usname, Filename)\} \wedge\\
&\qquad PublicFiles' = PublicFiles \cup \{(usname, Filename)\})\\
&\quad \wedge\ NormalUsers' = NormalUsers\\
&\quad \wedge\ SystemUsers' = SystemUsers\\
&\quad \wedge\ FileStore' = FileStore\\
&\quad \wedge\ ArchiveFiles' = ArchiveFiles)
\end{aligned}$$

The operation of this command depends on the category of user.

> 8. The SET-PRIVATE command sets the state of an existing file to be private. The command can be used by both normal and system users. If a system user types the command, then any file can be marked private. If a normal user types the command, then only a file owned by that user can be made private.

The effect of the SET-PRIVATE command can now be specified.

$$
\begin{aligned}
& command = SET - PRIVATE \Rightarrow \\
& \quad (user \in NormalUsers \Rightarrow \\
& \qquad (PrivateFiles' = PrivateFiles \cup \{(user, FileName)\} \wedge \\
& \qquad PublicFiles' = PublicFiles \setminus \{(user, FileName)\}) \\
& \quad \wedge\ user \in SystemUsers \Rightarrow \\
& \qquad (PublicFiles' = PublicFiles \setminus \{(usname, FileName)\} \wedge \\
& \qquad PrivateFiles' = PrivateFiles \cup \{(usname, FileName)\}) \\
& \quad \wedge\ NormalUsers' = NormalUsers \\
& \quad \wedge\ SystemUsers' = SystemUsers \\
& \quad \wedge\ FileStore' = FileStore \\
& \quad \wedge\ ArchiveFiles' = ArchiveFiles)
\end{aligned}
$$

Again the operation of this command depends on the category of user.

> 9. The ARCHIVE command marks a file for archiving. At certain times of the working week a program known as an archiver is executed. The archiver removes files which are marked for archiving from the filestore and writes them to magnetic tape. This command can only be used by a system user.

The effect of the ARCHIVE command can now be specified.

$$
\begin{aligned}
& command = ARCHIVE \Rightarrow \\
& \quad (ArchiveFiles' = ArchiveFiles \cup \{(usname, FileName)\} \\
& \quad \wedge\ NormalUsers' = NormalUsers \\
& \quad \wedge\ SystemUsers' = SystemUsers \\
& \quad \wedge\ PrivateFiles' = PrivateFiles \\
& \quad \wedge\ PublicFiles' = PublicFiles)
\end{aligned}
$$

Only the archive files are affected by this command.

> 10. The DE-ARCHIVE command unmarks a file which has already been marked for archiving.

The effect of the DE-ARCHIVE command is

$$
\begin{aligned}
& command = DE - ARCHIVE \Rightarrow \\
& \quad (ArchiveFiles' = ArchiveFiles \setminus \{(usname, FileName)\} \\
& \quad \wedge\ NormalUsers' = NormalUsers \\
& \quad \wedge\ SystemUsers' = SystemUsers \\
& \quad \wedge\ PrivateFiles' = PrivateFiles \\
& \quad \wedge\ PublicFiles' = PublicFiles)
\end{aligned}
$$

Worked example 5.8 During the operation of the filing system a number of users asked for extra commands. One popular request was for a command which displayed the status of a file, i.e. whether it was private or public. The specification for this command was

> The STATUS command displays the status of a file in the file system. It can be used by both system users and by normal users. Any file can be examined by this command.

Write a specification which gives the effect of this command; do not worry about errors. Assume the existence of a predicate *DisplayStatus* that has a file name parameter and an output device parameter. The predicate is true when the status of its first parameter is displayed on its second parameter. ← *

Solution The specification which describes the effect of this command is

$$\begin{array}{l} command = STATUS \Rightarrow \\ \quad (DisplayStatus(UserDevice, FileName) \wedge unchanged) \end{array}$$

Notice that the user has to specify a *usname* since any file in the file store can be examined.

■

The example in this section is relatively realistic. To be totally realistic it would have to include details about directory structure and access rules for public and private files. Nevertheless, it contains a substantial part of the functionality expected in a filing system.

5.6 Further reading

[Stoll, 1961] is an excellent introduction to set theory, even though it was written thirty years ago. It's the type of book every mathematician should aim to write. [Kolman and Busby, 1987] is a very good introduction to discrete mathematics written for computer scientists. It certainly is one of the best-produced mathematics books that I know of.

* In answering worked example 5.8, you should assume the existence of a predicate - unchanged - which states that the stored data of the configuration management system is not altered.

6

Relations

Aims

- To describe the nature of relations.
- To show that relations model associations between objects.
- To describe a number of relational operators.

6.1 Relations as sets of ordered pairs

A **relation** is one of the most important concepts in pure mathematics. It is a set of ordered pairs. An **ordered pair** is a pair of items which have the property

$$(x, y) = (u, v) \equiv (x = u) \land (y = v)$$

i.e. two ordered pairs are equal *only when* their first elements are equal and their second elements are equal. The formal definition of a relation is

$$\{ a : A; \ b : B \mid p(a, b) \bullet (a, b) \}$$

where p is a predicate which relates a and b and defines members of the relation. For example, the set

$$\{ a, b : \mathbb{N} \mid a + b = 4 \bullet (a, b) \}$$

defines the relation

$$\{(0,4), (1,3), (2,2), (3,1), (4,0)\}$$

Note that since ordered pairs are equal only if their corresponding elements are equal the relation contains both (0,4), (1,3) and (4,0), (3,1).

A relation is used to express the fact that there is a connection between the elements that make up an ordered pair. In the constructive specification of a relation it is the predicate which defines this connection. For example, the predicate $a + b = 4$ used in the constructive specification on page 114 expresses the fact that the elements of each ordered pair that make up the relation always add up to 4.

Relations are normally named. Naming a relation is equivalent to naming the set of pairs that make up the relation. Thus, the relation *eqless*

$$\{ x, y : \mathbb{N} \mid x = y \land x < 4 \}$$

can be written as either the constructive specification

$$eqless = \{ x, y : \mathbb{N} \mid x = y \land x < 4 \}$$

or by enumeration

$$eqless = \{(0,0), (1,1), (2,2), (3,3)\}$$

The fact that an ordered pair (x, y) is contained in a relation can be written as $(x, y) \in R$ or $R(x, y)$. However, it is customary to write this predicate as xRy. When the ordered pairs that make up a relation are extracted from two sets A and B it is usual to refer to the relation as being over A and B, or as being on A and B, or as being over $A \times B$. This book will adopt the last convention. Thus, the relation *eqless* above is over $\mathbb{N} \times \mathbb{N}$.

Relations are not quite a new topic in this book. In Chapter 4 predicates such as *connected* and *uses* were employed to express the fact that there was a relation between two objects. It can now be seen that this was just another way of writing relations.

Worked example 6.1 Write down the elements of the relation

$$peoplen = \{ p : person;\ n : \mathbb{N} \mid n < 3 \bullet (p, n) \}$$

where $person = \{Roberts, Jones, Monroe\}$. Examine the following predicates

(i) $(Timms, 2) \in peoplen$
(ii) $(Williams, 2) \notin peoplen$
(iii) $Roberts\ peoplen\ 3$
(iv) $Jones\ peoplen\ 2$
(v) $Jones\ peoplen\ 4$

Which of them do not hold?

Solution The relation *peoplen* is the set of all ordered pairs whose first element is taken from *person* and whose second element is taken from $\mathbb{N}$. Since the second element will always be less than 3, the contents of the relation will be

$$\{(Roberts, 0), (Roberts, 1), (Roberts, 2), (Jones, 0), (Jones, 1), (Jones, 2), (Monroe, 0), (Monroe, 1), (Monroe, 2)\}$$

Hence (ii) and (iv) are true and (i), (iii), and (v) are all false.

■

Exercise 6.1
If *files* = {*archivedir*, *newlog*} and *sys* = {*archivedir*} write down the elements of the following files

(i) $\{ n : \mathbb{N};\ fil : files \mid 2 * n < 9 \bullet (n, fil) \}$
(ii) $\{ n : \mathbb{N};\ fil : files \mid file \in sys \land n < 4 \bullet (n, fil) \}$
(iii) $\{ n : \mathbb{N};\ fil : files \mid file \in sys \land n = 2 \bullet (n, fil) \}$
(iv) $\{ n : \mathbb{N};\ fil : files \mid n < 0 \bullet (n, fil) \}$
(v) $\{ fil : files;\ n : \mathbb{N} \mid n > 2 \land n < 5 \bullet (n^2, fil) \}$
(vi) $\{ fil : files;\ n : \mathbb{N} \mid n = 3 \land fil = newlog \bullet (n^3, fil) \}$

■

You will already be familiar with relations from numerical mathematics. For example, the less than relation, $<$, for natural numbers can be written as

$$\{ a, b : \mathbb{N} \mid \exists k : \mathbb{N}_1 \bullet a + k = b \}$$

where $\mathbb{N}_1$ is the set of natural numbers excluding zero. The elements of this infinite set are

$$\{(0,1), (0,2), (0,3), \ldots, (1,2), (1,3), (1,4), \ldots, (2,3), (2,4), (2,5), \ldots\}$$

Worked example 6.2 Why is $\mathbb{N}_1$ used in the definition of $<$?

Solution The definition of $<$ states that two elements are related if there is a natural number in $\mathbb{N}_1$ which, when added to the first element of the pair, gives the second element of the pair. If $\mathbb{N}$ were used, then elements such as (2,2) and (19,19) would be part of the relation $<$ since 0 added to the first element would give the second element. To exclude these possibilities $\mathbb{N}_1$ has to be used.

■

Relations, as their name suggests, describe a relationship between the elements. You have already seen an example of a relation in Section 5.5 where the file store of a computer was modelled as a relation over file names and user names, where the predicate which described the relation was the fact that a user owned a file. It is this property of relations that makes them useful for modelling computer systems.

Some examples of fragments of specifications which imply a relation exists between two objects are shown below.

> Privileged users are those users who have a priority of 7 or more.
>
> Each stock item will be associated with the number in stock and the minimum level below which the item must not fall before a back order is placed.
>
> The stock data base system should contain information which allows a query clerk to discover which division has ordered a particular shipment of parts.

The first fragment expresses the fact that there will be a relation over users and priorities. The second fragment embodies two relations: the first is over stock items and natural numbers and expresses the connection between a stock item and the minimum re-order level. The third fragment embodies a relation over divisions and orders.

The elements of a relation can be ordered pairs whose elements are sets, for example, if

$$\mathit{files} = \{\mathit{new}, \mathit{old}, \mathit{archive}, \mathit{summary}, \mathit{tax}\}$$

and

$$\mathit{users} = \{\mathit{Jones}, \mathit{Roberts}, \mathit{Wilson}\}$$

the relation $\mathit{CanAccess}$

$$\{(\mathit{Jones}, \{\mathit{new}\}), (\mathit{Roberts}, \{\mathit{new}, \mathit{old}, \mathit{summary}\}), (\mathit{Wilson}, \{\mathit{tax}\})\}$$

is an example of a relation over $\mathit{users} \times \mathbb{P}\,\mathit{files}$ which may describe those files which a particular user can access.

A number of operators are defined for relations. The domain operator dom has one operand; its value is the set whose members are the left-hand elements of the pairs in a relation. Thus, if owns is a relation over $\mathit{users} \times \mathit{files}$ and its current value is

$$\{(\mathit{Jones}, \mathit{tax}), (\mathit{Jones}, \mathit{new}), (\mathit{Roberts}, \mathit{summary}), (\mathit{Wilson}, \mathit{archive})\}$$

then $\mathrm{dom}\,\mathit{owns}$ is

$$\{\mathit{Jones}, \mathit{Roberts}, \mathit{Wilson}\}$$

Formally dom is defined as

$$\operatorname{dom} A = \{\ t_1 : T_1 \mid \exists\, t_2 : T_2 \bullet t_1 A t_2\ \}$$

where A is a relation over $T_1 \times T_2$.

The ran operator is similar to dom. It returns with the set of right-hand elements in a relation. Thus, if *UsesComputer* is a relation over *users* × *computers* where

$$computers = \{\, VAX780, Sequent, DRS800\}$$

and the current value of *UsesComputer* is

$$\{(Jones, VAX780), (Roberts, Sequent), (Wilson, VAX780)\}$$

then ran *UsesComputer* is

$$\{\, VAX780, Sequent\}$$

Worked example 6.3 The users of a computer operating system are modelled by means of two sets *NormalUsers* and *PrivilegedUsers*. The association between users and the files that they own is also modelled by means of a relation *owns* over *users* and *files*. Write down the expression which represents the set of files owned by privileged users.

Solution The specification that describes this set is shown below.

$$\operatorname{ran}\{\ u : users;\ f : files \mid u \in PrivilegedUsers \land (u, f) \in owns\}$$

The specification forms the relation which contains privileged users and files.

■

Worked example 6.4 Formally define the ran operator.

Solution The definition is

$$\operatorname{ran} A = \{\ t_2 : T_2 \mid \exists\, t_1 : T_1 \bullet t_1 A t_2\ \}$$

where the relation A is over $T_1 \times T_2$ where T_1 and T_2 are any types. All the definition states is that the range of a relation is the set of t_2's which are related to a t_1 in A.

■

Exercise 6.2
Suppose *owns* and *CanRead* are relations over $users \times files$ and their current values are

$$owns = \{(Roberts, archive), (Wilson, tax), (Roberts, summary), (Jones, old)\}$$

$$CanRead = \{(Roberts, archive), (Wilson, archive), (Wilson, tax), (Jones, tax), (Roberts, summary), (Jones, old), (Jones, archive)\}$$

Indicate which of the following predicates are true and which are false.

(i) $Roberts\ owns\ archive$
(ii) $\neg\ (Roberts\ owns\ summary)$
(iii) $(Jones, tax) \in owns \wedge (Roberts, archive) \in CanRead$
(iv) $owns \subset CanRead$
(v) $\# owns = 7$
(vi) $\mathrm{dom}\ owns = \mathrm{dom}\ CanRead$
(vii) $Thomas \in owns \vee Ince \in \mathrm{dom}\ CanRead$
(viii) $\mathrm{dom}\ owns \cap \mathrm{dom}\ CanRead \neq \{Wilson, Timms\}$
(ix) $\mathrm{dom}\ owns \cap \{Timms\} = \varnothing$
(x) $\mathrm{dom}\ owns \cap \{Timms\} \neq \{Wilson, Timms\}$
(xi) $\# CanRead > 7$
(xii) $\{(Roberts, archive)\} \cup \{(Wilson, tax)\} \subset CanRead$
(xiii) $Roberts\ owns\ archive \wedge Roberts\ CanRead\ archive$
(xiv) $Roberts\ owns\ archive \wedge (Roberts, archive) \in CanRead$

■

A further useful operator is the **inverse operator**. It has one operand which is a relation. It reverses the elements of the pairs of the relation it operates on. The operator is written by writing -1 as a superscript above the relation whose inverse is to be expressed. Thus, if *owns* is a relation over $users \times computers$ and is currently

$$\{(Wilson, SUN1), (Jones, SUN2), (Timms, Apollo1)\}$$

then $owns^{-1}$ is

$$\{(SUN1, Wilson), (SUN2, Jones), (Apollo1, Timms)\}$$

which is just the individual pairs reversed.

Worked example 6.5 Formally define the inverse operator.

Solution The inverse operator takes each ordered pair in a relation and reverses the elements. This can be expressed as

$$\{\ t_1 : T_1;\ t_2 : T_2 \mid (t_1, t_2) \in R \bullet (t_2, t_1)\ \}$$

where R is a relation over any two arbitrary types T_1 and T_2.
■

Exercise 6.3
size is a relation over $FileNames \times \mathbb{N}$. Its current value is

$$\{(archive, 273), (tax, 123), (lister, 459), (newarch, 450)\}$$

Which of the following predicates are true and which are false?

(i) $\operatorname{dom} size^{-1} = \{archive, lister, tax, newarch\}$
(ii) $\operatorname{dom} size^{-1} = \{273, 123, 459, 450\}$
(iii) $\# \operatorname{ran} size = 4$
(iv) $\# \operatorname{ran} size = \# \operatorname{dom} size$
(v) $\operatorname{dom}\{ fil : FileNames;\ n : \mathbb{N} \mid fil\ size\ n \land n < 400 \bullet (fil, n) \} = \{lister\}$
(vi) $\operatorname{dom}\{ fil : FileNames;\ n : \mathbb{N} \mid fil\ size\ n \land n < 400 \bullet (fil, n) \} = \{(273, archive), (123, tax)\}$
(vii) $(size^{-1})^{-1} = size$
(viii) $\operatorname{dom}\{ fil : FileNames;\ n : \mathbb{N} \mid fil\ size\ n \land n > 500 \bullet (fil, n) \} = \varnothing$

■

6.2 Relation composition

Relations can be combined by an operation known as **composition**. As an example of this consider the two relations *HasQueue* and *ContainsTrans*. The first relation is over $computers \times queues$ while the second is over $queues \times transactions$. *HasQueue* describes the fact that a particular computer has a series of queues associated with it; *ContainsTrans* describes the fact that a particular queue contains a transaction that is to be processed. Given these relations, it may be necessary to construct a relation *HasTrans* over $computers \times transactions$ which describes the fact that a computer is waiting to process a series of transactions. Thus, if the current value of *HasQueue* is

$$\{(VAXA, q1), (VAXB, q2), (SUNA, q3), (SUNB, q4)\}$$

and the current value of *ContainsTrans* is

$$\{(q1, upd1), (q1, upd3), (q3, read1), (q4, read2), (q3, read2), (q2, read4), (q2, read5), (q1, tranup)\}$$

then the relation *HasTrans* would be

$$\{(VAXA, upd1), (VAXA, upd3), (VAXA, tranup), (VAXB, read4), (VAXB, read5), (SUNA, read1), (SUNA, read2), (SUNB, read2)\}$$

HasTrans has been formed by taking the first elements of the pairs in *HasQueue* and combining them with the second elements of *ContainsTrans* whose first elements match the second elements of *HASQueue*. The operator used to indicate composition is the semicolon $⨾$. Thus, in the example above, the composition of *ContainsTrans* and *HasQueue* is written as

$$HasTrans = HasQueue \mathbin{⨾} ContainsTrans$$

It is important to realize that in the expression

$$relation_1 \mathbin{⨾} relation_2$$

the type of the range of $relation_1$ must match the type of the domain of $relation_2$. Thus,

$$\{(1,17),(2,17),(8,13),(11,47)\} \mathbin{⨾} \{(file1, Jones),(file2, Wilson)\}$$

is not defined since the range of the first relation contains the natural numbers while the domain of the second relation contains files. The formal definition of relation composition follows. It is defined for a relation R_1 over $T_1 \times T_2$ and R_2 over $T_2 \times T_3$ where T_1, T_2, and T_3 are any arbitrary type:

$$R_1 \mathbin{⨾} R_2 = \{\ t_1 : T_1;\ t_3 : T_3 \mid \exists\, t_2 : T_2 \bullet (t_1, t_2) \in R_1 \wedge (t_2, t_3) \in R_2\ \}$$

It states that the composition of two relations is formed by first finding any two pairs such that the second element of the first pair matches the first element of the second pair; a pair is then formed from the first element of the first pair and the second element of the second pair.

Exercise 6.4

computers is a set of computers whose current value is

$$\{VAXA, SUNB, SUNC\}$$

lines is a set of communication lines whose current value is

$$\{comm1, comm3, comm5\}$$

and *conn* is a relation over *computers* $\times$ *lines* which models the fact that a computer is connected to a communications line. Its current value is

$$\{(VAXA, comm1), (SUNB, comm3)\}$$

Write down the value of the following expressions:

(i) $(computers \times lines) \fatsemi \{(comm1, on), (comm3, off)\}$
(ii) $conn \fatsemi conn$
(iii) $conn \fatsemi conn^{-1}$
(iv) $conn^{-1} \fatsemi conn$
(v) $(computers \times lines)^{-1}$
(vi) $conn^{-1} \fatsemi (computers \times lines)$
(vii) $conn \subseteq computers \times lines$
(viii) $\{(VAXA, comm2)\} \subset conn$

∎

A relation can be composed with itself. As an example of this consider the relation *ReceivesData* which models the fact that one monitor receives data from another and is over $monitors \times monitors$. If its current value is

$$\{(mon1, mon3), (mon3, mon2), (mon4, mon2), (mon2, mon1)\}$$

then the composition of *ReceivesData* with itself is

$$\{(mon1, mon2), (mon3, mon1), (mon4, mon1), (mon2, mon3)\}$$

The new relation $ReceivesData \fatsemi ReceivesData$ so formed describes the fact that a monitor can receive data from another monitor via another; thus,

$$(mon1, mon2) \in ReceivesData \fatsemi ReceivesData$$

asserts the fact that $mon2$ receives data from $mon1$ via another monitor which, in this case is $mon3$.

Worked example 6.6 How might you express the fact that a monitor, say $mona$, receives data from another monitor, say $monb$, via three other monitors?

Solution The answer is that

$$(mona, monb) \in$$
$$(ReceivesData \fatsemi ReceivesData \fatsemi ReceivesData \fatsemi ReceivesData)$$

As will be seen later a relation can be composed many times with itself.

∎

It is important to point out that a relation can be composed with itself only if its domain is the same as the type of its range. Such relations are termed **homogeneous**. In general, a relation can be composed with itself any number of times. Thus, one can write

$$IsConnected \fatsemi IsConnected$$

for a two-fold composition of the homogeneous relation *IsConnected*, and

$$IsConnected \fatsemi IsConnected \fatsemi IsConnected$$

for a three-fold composition. For an n-fold composition it is customary to use a power notation involving superscripts. For example, the four-fold composition of the relation *IsConnected* would be written as

$$IsConnected^4$$

6.3 The identity relation

An important relation used in specifications is the **identity relation**. It is formally defined as

$$\mathrm{id}\, A = \{\ a : A \bullet (a, a)\ \}$$

where A is any set whose elements are of an arbitrary type. Thus, $\mathrm{id}\{1, 3, 5, 9\}$ is

$$\{(1,1), (3,3), (5,5), (9,9)\}$$

and $\mathrm{id}\{Wilson, Wallis\}$ is

$$\{(Wilson, Wilson), (Wallis, Wallis)\}$$

Worked example 6.7 The relation *OwnsFile* is over $users \times FileNames$. Its current value is

$$\{(Jones, upd), (Wilson, tax1), (Jones, newupd), (Harris, newloc)\}$$

What is the value of the expression

$$(\mathrm{id}\{Jones, Wilson\}) \fatsemi OwnsFiles?$$

Can you express the relationship between this expression and *OwnsFiles* in natural language?

Solution The value of the expression is

$$\{(Jones, Jones), (Wilson, Wilson)\} \fatsemi$$
$$\{(Jones, upd), (Wilson, tax1), (Jones, newupd), (Harris, newloc)\}$$

which gives

$$\{(Jones, upd), (Wilson, tax1), (Jones, newupd)\}$$

The effect of the operations is to form the relation between Jones and Wilson and the files which are owned by them.

■

Exercise 6.5
The operation of a chemical plant computer system can be modelled using the sets: *reactors*, *monitors*, and *computers* where

$$reactors = \{cracker1, cracker2, cracker3, distill1, distill2\}$$

$$monitors = \{mon1, mon2, mon3, mon4, mon7, mon8, mon9\}$$

$$computers = \{VAXA, VAXB\}$$

The fact that a monitor is connected to a reactor is recorded in a relation *ReactorConnectedTo* which is over $reactors \times monitors$. The fact that a computer is connected to a monitor is recorded in the relation *ComputerConnectedTo* which is over $computers \times monitors$. If the current value of *ReactorConnectedTo* is

$$\{(cracker1, mon1), (cracker2, mon2), (cracker2, mon3), (cracker3, mon1), (cracker3, mon2), (distill1, mon7), (distill2, mon8), (distill2, mon9)\}$$

and the current value of *ComputerConnectedTo* is

$$\{(VAXA, mon1), (VAXA, mon2), (VAXB, mon2), (VAXB, mon3), (VAXB, mon7), (VAXB, mon8), (VAXB, mon9)\}$$

then what is the value of the following?

(i) $\mathrm{id}\ reactors$
(ii) $\mathrm{id}(reactors^{-1})$
(iii) $(\mathrm{id}\ reactors)^{-1}$
(iv) $(\mathrm{id}\ computers) \mathbin{;} ComputerConnectedTo$
(v) $(\mathrm{id}\{VAXA\}) \mathbin{;} ComputerConnectedTo$
(vi) $\mathrm{ran}((\mathrm{id}\{VAXA\}) \mathbin{;} ComputerConnectedTo)$
(vii) $\mathrm{id}(\{cracker1, cracker2\}) \mathbin{;} ReactorConnectedTo$
(viii) $\mathrm{ran}(\mathrm{id}(\{cracker3\}) \mathbin{;} ReactorConnectedTo)$
(ix) $\mathrm{id}\ \mathrm{dom}(ReactorConnectedTo)$
(x) $\mathrm{id}(\mathrm{dom}(ReactorConnectedTo) \setminus \{cracker1, cracker2\})$
(xi) $\mathrm{ran}(\mathrm{id}(\{VAXA\}) \mathbin{;} ComputerConnectedTo) \setminus \{mon1, mon2, mon9\}$

■

Worked example 6.8 The relation *updates* is a relation over $units \times variables$. It describes the fact that a program unit (procedure or function) updates a particular global variable. Write down the relation *SameUpdate* over $units \times units$ which describes the fact that two units are related if they update the same variable. The relation should be expressed in terms of *updates* using relational composition.

Solution The relation *SameUpdate* can be expressed as

$$SameUpdate = (updates \mathbin{;} updates^{-1}) \setminus \mathrm{id}\ updates$$

The identity function is required since $updates \mathbin{;} updates^{-1}$ will contain pairs containing identical elements made up of the first elements of the pairs in *updates*.
■

6.4 Relation restriction

In system specifications there is a need for relations to be restricted over their domain or range. For example, the specification extract

> The effect of the TEMP command is to display the current temperature of the monitors connected to the reactors which are specified in the command.

implies that if the connection between monitors and reactors is modelled by a relation, then the part of the specification for this command involves restricting the relation to include only those pairs which reference the reactors specified in the command.

Since restriction occurs time and time again in specifications, a number of operators can be defined which describe various forms of this restriction. The first of these operators $\lhd$ restricts the domain. It has two operands: the first operand is a set; the second operand is a relation. It forms a subset of the second operand which only contains pairs whose first elements are contained in the set. For example, if *IsConnected* is a relation over $reactors \times monitors$ and has the value

$$\{(reactor1, mon1), (reactor2, mon2), (cracker1, mon2),\\ (cracker2, mon1), (cracker2, mon3), (cracker3, mon4),\\ (distill1, mon4), (distill1, mon1), (distill2, mon1)\}$$

then

$$\{reactor1, reactor2\} \lhd IsConnected$$

is

$$\{(reactor1, mon1), (reactor2, mon2)\}$$

and

$$\{distill3\} \lhd IsConnected$$

is the empty set $\varnothing$. A similar operator is $\rhd$ which restricts the range of a relation. For example, the value of

$$IsConnected \rhd \{mon1, mon4\}$$

is

$$\{(reactor1, mon1), (cracker2, mon1), (cracker3, mon4),$$
$$(distill1, mon4), (distill1, mon1), (distill2, mon1)\}$$

Both these operators can be defined in terms of the operators previously described

$$S \lhd R = (\mathrm{id}\, S) \mathbin{;} R$$
$$R \rhd T = R \mathbin{;} (\mathrm{id}\, T)$$

where R is a relation over two arbitrary types T_1 and T_2, S is a subset of T_1, and T is a subset of T_2. Two further restriction operators can be defined. The operator $⩤$ is similar to $\lhd$ except that the subset over which the restriction is to hold includes those elements *not* specified as the first operand. In order to illustrate the use of this operator consider the relation *AccessFile* over *Users* $\times$ *files* which models the fact that a user can access a number of files where

$$users = \{Williams, Thomas, Jones, Ross, Timms\}$$
$$files = \{upd3, upd4, upd5, tax1, tax2, tax3\}$$

and where the current value of *AccessFiles* is

$$\{(Williams, upd3), (Thomas, upd4), (Jones, upd4), (Ross, tax1),$$
$$(Ross, tax2), (Timms, upd3), (Timms, tax3)\}$$

Then the value of $\{Williams, Jones\}$ $⩤$ *AccessFile* is

$$\{(Thomas, upd4), (Ross, tax1), (Ross, tax2),$$
$$(Timms, upd3), (Timms, tax3)\}$$

The operator $⩥$ is similar to $\rhd$ in its action. It restricts the range of its operand to those elements which are not contained in its second operand. For example, the value of

$$AccessFiles ⩥ \{upd3, tax3\}$$

is

$$\{(Thomas, upd4), (Jones, upd4), (Ross, tax1), (Ross, tax2)\}$$

Again these operators can be defined in terms of previously defined operators

$$S \lessdot R = (T_1 \setminus S) \lhd R$$

$$R \gtrdot T = R \rhd (T_2 \setminus T)$$

where R is a relation over $T_1 \times T_2$, S is a subset of T_1, and T is a subset of T_2.

One operator which is related to the restriction operators is the relational **override** operator $\oplus$. This is often used in specifications where a number of relations are used and where one relation is very much like another except for some pairs. Typical examples of this which occur in specifications are:

> The effect of the POOL operation is to change those files in the file store which are specified in the command so that they are owned by the system manager.
>
> The effect of the UPDATE command is to replace existing accounts by new accounts. These new accounts may have the same account number as those which are replaced but will have a different account holder.

The formal definition of $\oplus$ is

$$R \oplus S = (\text{dom}\, S \lessdot R) \cup S$$

where S and R are relations over two arbitrary types T_1 and T_2. The effect of the $\oplus$ operator is to replace those pairs in R whose first elements are in the domain of S by the elements of S. Thus,

$$\{(1, file1), (3, upd), (5, tax), (7, new), (8, upd)\} \oplus$$
$$\{(1, newtax), (7, oldtax)\}$$

is equal to

$$\{(1, newtax), (3, upd), (5, tax), (7, oldtax), (8, upd)\}$$

A typical application of $\oplus$ might occur in a data base application for banking where the fact that an account holder has a numbered account is modelled by means of the relation *HasAccount* which is over $\mathbb{N} \times names$. At times during the day a new account has to be added to *HasAccount* and, occasionally, a new account is assigned to an existing account holder. If these are held in the relation *update* which is over $\mathbb{N} \times names$ then the effect of updating *HasAccount* is

$$HasAccount' = HasAccount \oplus update$$

Note that this operation, as well as introducing new accounts, reassigns new account holders to existing accounts. For example, if the current value of *HasAccount* is

$$\{(1173, Jones), (1148, Thomas), (1176, Jones), (493, Wilson)\}$$

and *update* is

$$\{(1148, Timms), (1922, Roberts), (1111, Wilson)\}$$

then the new value of *HasAccount* would be

$$\{(1173, Jones), (1148, Timms), (1176, Jones),\\ (493, Wilson), (1922, Roberts), (1111, Wilson)\}$$

Worked example 6.9 What would be the value of

$$\varnothing \oplus \{(Williams, 125), (Horton, 123)\}?$$

It may be worth looking at the formal definition of the $\oplus$ operator before constructing an answer.

Solution The value would be

$$\{(Williams, 125), (Horton, 123)\}$$

All that would happen would be that the empty set would be unioned with the set containing the two pairs.

■

Another useful operator is the **relational image operator**. The image of a set S through a relation R is the set of elements which are second elements of pairs contained in R whose first elements are in S. Formally, the relational image operator $(\!|\ |\!)$ is defined as

$$R(\!|S|\!) = \{\ t_2 : T_2 \mid \exists\, t_1 : S \bullet t_1 R t_2\ \}$$

where R is over $T_1 \times T_2$ and S is a subset of T_1.

For example, if the set $HasAccess$ is

$$\{(Jones, tax2), (Roberts, tax3), (Roberts, upd), (Thomas, newupd),\\ (Thomas, tax2), (Wilson, tax3), (Timms, trans)\}$$

and if the set $FileQuery$ was

$$\{Jones, Roberts\}$$

then $HasAccess(\!|FileQuery|\!)$ would be

$$\{tax2, tax3, upd\}$$

and $HasAccess(\!|\{Wilkinson\}|\!)$ would be the empty set $\varnothing$.

Exercise 6.6
A data base application for a wholesaler can be modelled by means of the relations *price*, *InStock*, and *supplies*. *Price* is a relation over $products \times 1\,..\,20$ which models the association between prices and stock items. *InStock* is a relation over $products \times \mathbb{N}$ which models the association between stock items and the current number in stock of a product. *Supplies* is a relation over $suppliers \times products$ which models the relation between a supplier and the product that is delivered by that supplier. If the current values of these relations are

$$price = \{(nut, 3), (bolt, 5), (screw, 1), (board, 17), (fastner, 12)\}$$

$$InStock = \{(nut, 500), (bolt, 2100), (screw, 45), (board, 0), (fastner, 500)\}$$

$$supplies = \{(Thomas, nut), (Thomas, bolt), (Wilks, bolt), (Wilks, screw),\\ (Wilks, board), (Wilks, fastner), (Rogers, board), (Rogers, fastener)\}$$

then what are the values of the following expressions? Also express in natural language the relations described by each expression.

(i) $\mathrm{id}\ price$
(ii) $\{nut, bolt\} \lhd price$
(iii) $\mathrm{dom}(price \rhd\!\!\!- 1\,..\,5)$
(iv) $\mathrm{dom}(price \rhd 1\,..\,5)$
(v) $\mathrm{ran}(price \rhd 1\,..\,5)$
(vi) $InStock \rhd \{0\}$
(vii) $\mathrm{dom}(supplies\,;(InStock \rhd \{0\}))$
(viii) $price \oplus \{(bolt, 6)\}$
(ix) $price \oplus \{(hanger, 2), (screw, 2)\}$
(x) $\mathrm{dom}(supplies\,;(price \rhd\!\!\!- 5\,..\,20))$
(xi) $supplies(\!|\{Robinson\}|\!)$
(xii) $supplies(\!|\{Rogers, Wilson\}|\!)$
(xiii) $((supplies\,;price) \rhd 1\,..\,10)(\!|\{Robinson, Rogers\}|\!)$

(xiv) $supplies \setminus (\{Thomas\} \lhd supplies)$
(xv) $\{Thomas\} \ntriangleleft supplies$
(xvi) $(InStock \rhd \{0\})(\!|\{nut, bolt\}|\!)$
(xvii) $\#\mathrm{dom}(supplies(\!|\{bolt\}|\!))$

■

Exercise 6.7
The file store of a small computer is modelled by the relations: *owns*, *size*, *HasUserStatus*, and *HasFileStatus*. *Owns* is a relation over $users \times files$ which models the fact that the user of a computer owns a particular file. *Size* is a relation over $files \times FileSizes$ which relates a file to its size in kbytes. *HasUserStatus* is a relation over $users \times UserStatus$ which relates a user to his or her status in the computer system and *HasFileStatus* is a relation over $files \times FileStatus$ which relates a file to its status, where

$$FileSizes = 1 \mathinner{..} 100$$

$$UserStatus = \{normal, superuser\}$$

$$FileStatus = \{read, write, delete\}$$

Write down expressions that are equivalent to the following natural language extracts taken from the system specification for the file access system. Use the relational operators described in this chapter.

(i) At most only three super users are allowed at one time.
(ii) The total number of users who own files should never exceed 15.
(iii) No user is allowed to own a file which has delete status and which is over 8 kbytes in size.
(iv) When the file name and name have been typed in the file size is increased by 1 kbyte.
(v) ... those files which do not belong to a normal user and which are under 6 kbytes long.
(vi) ... those files which do not belong to a super user and have write status.
(vii) ... those users who own files greater than or equal to 7 kbytes in size.

■

6.5 The transitive closure of a relation

An important operator is known as the **transitive closure** operator. However, before defining what exactly is meant by this, it is first necessary to formally define an operator which is used in its definition and which was defined informally earlier in this section. The nth iterate of a homogeneous relation is its n-fold composition with itself. It is written with a superscript over the relation. For example,

$$IsInSys^4$$

is the four-fold composition of $IsInSys$, i.e.

$$IsInSys \mathbin{;} IsInSys \mathbin{;} IsInSys \mathbin{;} IsInSys$$

The formal definition of the operator is shown below. If R is a homogeneous relation on $T \times T$ where T is any type, then

$$R^0 = \mathrm{id}\ T$$

and

$$R^n = R \mathbin{;} R^{n-1}$$

Thus, if a relation $LowerAddress$ on $programs \times programs$ has a value

$$\{(us1, us2x), (us2x, us3x), (us3x, us9b), (us9b, us7y)\}$$

where $programs$ is

$$\{us1, us2x, us3x, us9b, us7y, us11x, us14r\}$$

then $LowerAddress^3$ will, from the definition of composition, be

$$\{(us1, us2x), (us2x, us3x), (us3x, us9b), (us9b, us7y)\} \mathbin{;} LowerAddress^2$$

This can be further expanded to

$$\{(us1, us2x), (us2x, us3x), (us3x, us9b), (us9b, us7y)\} \mathbin{;}$$
$$\{(us1, us2x), (us2x, us3x), (us3x, us9b), (us9b, us7y)\} \mathbin{;} LowerAddress$$

This can be further simplified using the definition of $\mathbin{;}$ to

$$\{(us1, us3x), (us2x, us9b), (us3x, us7y)\} \mathbin{;} LowerAddress$$

which can then be written as

$$\{(us1, us3x), (us2x, us9b), (us3x, us7y)\} \mathbin{;}$$
$$\{(us1, us2x), (us2x, us3x), (us3x, us9b), (us9b, us7y)\}$$

which, when simplified, is

$$\{(us1, us9b), (us2x, us7y)\}$$

Given this definition of composition the transitive closure of a relation can be defined. There are in fact two transitive closures: the **reflexive transitive closure** and the **non-reflexive transitive closure**.

The reflexive transitive closure of a homogeneous relation is the union of all its iterates; its non-reflexive closure is the union of all its iterates except for its zeroth. The reflexive transitive closure of a relation R, written R^*, is therefore

$$R^0 \cup R^1 \cup R^2 \cup R^3 \ldots \cup R^n$$

The non-reflexive transitive closure of a homogeneous relation R, written R^+, is therefore

$$R^1 \cup R^2 \cup R^3 \ldots \cup R^n$$

Worked example 6.10 If *peripherals* is the set

$$\{reader1, reader2, lp1, lp2, lp3\}$$

and *IsAlternative* is a homogeneous relation over *peripherals* × *peripherals* whose current value is

$$\{(reader1, reader2), (lp1, lp2), (lp2, lp3), (lp3, lp1)\}$$

then what is the value of $alternative^+$?

Solution *IsAlternative* is

$$\{(reader1, reader2), (lp1, lp2), (lp2, lp3), (lp3, lp1)\}$$

$IsAlternative^2$ is

$$\{(lp1, lp3), (lp2, lp1), (lp3, lp2)\}$$

$IsAlternative^4$ is

$$\{(lp1, lp2), (lp2, lp3), (lp3, lp1)\}$$

also

$$IsAlternative^5 = IsAlternative^2$$

and

$$IsAlternative^6 = IsAlternative^3$$

Therefore $IsAlternative^+$ is

$$\{(reader1, reader2), (lp1, lp2), (lp2, lp3), (lp3, lp1), (lp1, lp3), (lp2, lp1), (lp3, lp2), (lp1, lp1), (lp2, lp2), (lp3, lp3)\}$$

■

As an example of the use of transitive closures consider a programming language such as Ada or Modula2 which supports modules. A large software developer will construct a system using these languages as a series of modules which are separately compiled. Each module will consist of a set of facilities such as procedures and functions and will use facilities supplied by other modules.

A module consists of a specification part which provides details of the facilities offered by the module, and an implementation part which is the program code that implements the facilities. If the specification part of a module is changed, then all the modules which use that changed module will need to be re-compiled.

If a module needs another module for its processing then this can be modelled by the homogeneous relation *NeedsModule* over *modules* × *modules*. Thus, if *ModuleA* requires facilities provided by *ModuleB* then

$$ModuleA\ NeedsModule\ ModuleB$$

is true.

If, as part of a language support system, a facility were provided which would enable a user to type in a series of modules and display which modules are needed for those typed modules, however indirectly, then the specification for this command would involve the non-reflexive transitive closure of *NeedsModule*. Thus, $NeedsModule^+$ is a relation which describes the dependencies (direct or indirect) between the developed modules in a software system.

Worked example 6.11 The relation *NextTo* is over *programs* × *programs*. It expresses the fact that programs in a computer system are located next to each other in main memory. Thus,

$$ProgramA\ NextTo\ ProgramB$$

is true when *ProgramA* is next to *ProgramB* with *ProgramB* occupying a higher address. The relation *HigherAddress* is over *programs* × *programs* and expresses the fact that one program occupies a higher address than another program. Express *HigherAddress* in terms of *NextTo*.

Solution The relationship is

$$HigherAddress = NextTo^{+}$$

the non-reflexive transitive closure.
■

6.6 Theorems involving relations

In the same way that theorems can be developed and proved in set theory and predicate calculus, a number of theorems can be derived and proved concerning relations.

If R_1 and R_2 are relations over $T_1 \times T_2$ and $T_2 \times T_3$ then

$$\vdash (R_1 \mathbin{;} R_2)^{-1} = R_2^{-1} \mathbin{;} R_1^{-1}$$

If R_1 is a relation over $T_1 \times T_2$, R_2 a relation over $T_2 \times T_3$, and R_3 a relation over $T_3 \times T_4$,

$$\vdash R_1 \mathbin{;} (R_2 \mathbin{;} R_3) = (R_1 \mathbin{;} R_2) \mathbin{;} R_3$$

If R is any relation over $T_1 \times T_2$ then

$$\vdash R \mathbin{;} (\mathrm{id}\ T_2) = R$$
$$\vdash (\mathrm{id}\ T_1) \mathbin{;} R = R$$

If R_1 and R_2 are relations over $T_1 \times T_2$ and R_3 a relation over $T_2 \times T_3$, then

$$\vdash (R_1 \cup R_2) \mathbin{;} R_3 = (R_1 \mathbin{;} R_3) \cup (R_2 \mathbin{;} R_3)$$
$$\vdash R_1 \subseteq R_2 \Rightarrow (R_1 \mathbin{;} R_3) \subseteq (R_2 \mathbin{;} R_3)$$

If R_1, R_2, and R_3 are relations over $T_1 \times T_2$ then

$$\vdash (R_1 \cup R_2)^{-1} = R_1^{-1} \cup R_2^{-1}$$
$$\vdash (R_1 \setminus R_2)^{-1} = R_1^{-1} \setminus R_2^{-1}$$
$$\vdash R_1 \oplus (R_2 \oplus R_3) = (R_1 \oplus R_2) \oplus R_3$$
$$\vdash R \oplus \varnothing = R$$
$$\vdash \varnothing \oplus R = R$$

All these theorems can be derived from the definitions of the relational operators presented in this chapter together with theorems from predicate calculus and set theory.

7

Functions and sequences

Aims

- To outline the nature of functions.
- To describe a particular type of function known as a sequence.
- To describe two small case studies, one using functions, the other using sequences.

7.1 Functions

A **function** is an important type of relation. It has the property that each element of its domain is associated with just one element of its range. Thus,

$$\{(1, \mathit{file}2), (3, \mathit{file}4), (7, \mathit{file}2), (6, \mathit{filetax}), (9, \mathit{fileupd})\}$$

is an example of a function while

$$\{(1, \mathit{file}2), (3, \mathit{file}5), (7, \mathit{file}3), (1, \mathit{file}5), (2, \mathit{file}4)\}$$

is not an example since 1 is associated with both the files *file*2 and *file*5. When a pair of elements occurs in a function it is said that the function **maps** the first element to the second element. Thus, in the function above 1 is mapped to *file*2 and 6 is mapped to *filetax*.

A **partial function** is a function whose domain is a proper subset of the set from which the first elements of its pair are taken. Thus, the function

$$\{(1,3), (4,9), (8,3)\}$$

over $\mathbb{N} \times \mathbb{N}$ is a partial function because its domain, $\{1,4,8\}$, is a proper subset of the natural numbers. Examples of partial functions abound in system specifications. For example,

> The main salesman file will hold data on the current performance of salesmen; this data will be retrieved by typing a unique salesman identity number.
>
> Each file in the file store will have a unique name. This is formed by concatenating the user name and the directory name. A file will be held in a non-overlapping area of the file store.
>
> Each entry in the process table consists of a process name and its location in store and corresponds to a currently functioning process. There will be no duplicate entries in the table.

The first paragraph describes a partial function whose domain is a subset of salesman numbers. Its range is a subset of performance data. The second paragraph describes a function whose domain is a subset of all user names concatenated with directory names; its range is a subset of possible addresses. Last, the third paragraph describes a function whose domain is a subset of possible process names and whose range is a subset of the natural numbers.

Worked example 7.1 Examine the system specification extract shown below.

> Each input device will have an entry in the peripheral table; the contents of the table entry represent the state of the device. The system will also store in the activity file each input device together with the names of those programs which currently have the input device allocated to them.

Identify any functions in the extract.

Solution First, there is a function which models the association between an input device and its state. The association between programs and devices which are allocated to them can either be modelled by means of a relation over $InputDevices \times programs$ or a function over $InputDevices \times \mathbb{P}\ programs$ ∎

A relation R over $T_1 \times T_2$ is a partial function if and only if

$$\forall t_1 : T_1;\ t_2, t_3 : T_2 \bullet (t_1 R t_2 \land t_1 R t_3) \Rightarrow t_2 = t_3$$

This states that in a partial function t_1 is mapped to both t_2 and t_3 only when both t_2 and t_3 are equal.

An important type of function is a **total function**. This is a function whose domain is equal to the set from which the first elements of its pairs are taken. For example, if the set $SysProgs$ is

$$\{archiver, editor, compilerA, compilerB, filer\}$$

and *location* is a function over $SysProgs \times \mathbb{N}$

$$\{(archiver, 12), (editor, 480), (compilerA, 903), (compilerB, 202), (filer, 17)\}$$

then *location* is a total function because its domain

$$\{archiver, editor, compilerA, compilerB, filer\}$$

is equal to $SysProgs$.

Formally, a function R over $T_1 \times T_2$ is total if

$$\forall t_1 : T_1;\ t_2, t_3 : T_2 \bullet \\ t_1 R t_2 \wedge t_1 R t_3 \Rightarrow t_2 = t_3 \wedge \operatorname{dom} R = T_1$$

Exercise 7.1
If

$$SysProg = \{archiver, editor, filer, compiler, linker\}$$

then classify the following functions

(i) $\{(archiver, 2), (editor, 3)\}$
(ii) $\{(archiver, 3), (editor, 3), (compiler, 3), (linker, 3), (filer, 3)\}$
(iii) $\{(archiver, 4), (editor, 4)\}$
(iv) $\{(archiver, 1)\}$
(v) $\{(archiver, 6), (compiler, 9), (editor, 12), (linker, 3), (filer, 84)\}$

■

An **injective function** is a function whose inverse is also a function. For example,

$$\{(1, 17), (2, 12), (3, 18), (4, 19)\}$$

is an injective function since its inverse

$$\{(17, 1), (12, 2), (18, 3), (19, 4)\}$$

is also a function, while

$$\{(8, 41), (9, 53), (10, 77), (11, 53)\}$$

is not injective because of the multiple occurrences of 53 as the second element in its pairs.

Injective functions can again be classified as partial or total. A function R over $T_1 \times T_2$ is a **partial injective function** if

$$\forall t_1 : T_1;\ t_2, t_3 : T_2 \bullet t_1 R t_2 \wedge t_1 R t_3 \Rightarrow t_2 = t_3 \wedge \\ \forall t_1, t_3 : T_1;\ t_2 : T_2 \bullet t_2 R^{-1} t_1 \wedge t_2 R^{-1} t_3 \Rightarrow t_1 = t_3$$

A function R over $T_1 \times T_2$ is a **total injective function** if

$$\forall t_1 : T_1;\ t_2, t_3 : T_2 \bullet t_1 R t_2 \wedge t_1 R t_3 \Rightarrow t_2 = t_3 \wedge$$
$$\forall t_1, t_3 : T_1;\ t_2 : T_2 \bullet t_2 R^{-1} t_1 \wedge t_2 R^{-1} t_3 \Rightarrow t_1 = t_3 \wedge$$
$$\operatorname{dom} R = T_1$$

Exercise 7.2
If *ProcessNames* is

$$\{process1, process2, process3, process4\}$$

and *range* is $1 \mathinner{..} 4$, then classify the following functions.

(i) $\{(1, process1)\}$
(ii) $\{(1, process1), (2, process1)\}$
(iii) $\{(1, process1), (2, process2), (3, process1), (4, process4)\}$
(iv) $\{(1, process1), (2, process2), (3, process3), (4, process4)\}$
(v) $\{(1, process1), (3, process1)\}$

∎

A function is said to be **surjective** if its range is the whole of the set from which the second elements of its pairs are taken. For example, if *computers* is the set

$$\{VAXA, VAXB, VAXC, SUNA\}$$

then the function

$$\{(1, VAXA), (3, VAXB), (7, VAXC), (9, VAXC), (11, SUNA)\}$$

is a surjective function over $\mathbb{N} \times computers$ since its range is

$$\{VAXA, VAXB, VAXC, SUNA\}$$

Again surjective functions can be total or partial. A function over $T_1 \times T_2$ is a **partial surjection** if

$$\forall t_1 : T_1;\ t_2, t_3 : T_2 \bullet t_1 R t_2 \wedge t_1 R t_3 \Rightarrow t_2 = t_3 \wedge \operatorname{ran} R = T_2$$

and a function R over $T_1 \times T_2$ is a **total surjection** if

$$\forall t_1 : T_1;\ t_2, t_3 : T_2 \bullet t_1 R t_2 \wedge t_1 R t_3 \Rightarrow t_2 = t_3 \wedge \operatorname{ran} R = T_2 \wedge$$
$$\operatorname{dom} R = T_1$$

Thus, if *monitors* is the set

$$\{mon1, mon2, mon3, mon4, mon5\}$$

and *reactors* is the set

$$\{reactor1, reactor2, reactor3\}$$

then

$$\{(mon1, reactor1), (mon2, reactor2), (mon3, reactor3),\\ (mon4, reactor3)\}$$

is a partial surjection and

$$\{(mon1, reactor1), (mon2, reactor2), (mon3, reactor2),\\ (mon4, reactor3), (mon5, reactor1)\}$$

is a total surjection.

A final important category of function is one which is both injective and surjective. These are known as **bijective functions**. A function R over $T_1 \times T_2$ is a partial bijective function if

$$\forall\, t_1 : T_1;\ t_2, t_3 : T_2 \bullet t_1 R t_2 \wedge t_1 R t_3 \Rightarrow t_2 = t_3 \wedge\\ \forall\, t_1, t_3 : T_1;\ t_2 : T_2 \bullet t_2 R^{-1} t_1 \wedge t_2 R^{-1} t_3 \Rightarrow t_1 = t_3 \wedge\\ \operatorname{ran} R = T_2$$

A function R over $T_1 \times T_2$ is a **total bijective function** if

$$\forall\, t_1 : T_1;\ t_2, t_3 : T_2 \bullet t_1 R t_2 \wedge t_1 R t_3 \Rightarrow t_2 = t_3 \wedge\\ \forall\, t_1, t_3 : T_1;\ t_2 : T_2 \bullet t_2 R^{-1} t_1 \wedge t_2 R^{-1} t_3 \Rightarrow t_1 = t_3 \wedge\\ \operatorname{ran} R = T_2 \wedge \operatorname{dom} R = T_1$$

An example of a total bijective function over

$$\{mon1, mon2, mon3\} \times \{reactor1, reactor2, reactor3\}$$

is

$$\{(mon1, reactor2), (mon2, reactor1), (mon3, reactor3)\}$$

Its inverse is a function, its domain is the set from which the first elements of its pairs are taken, and its range is the set from which the second elements of its pairs are taken.

Exercise 7.3
Classify the following finite functions. Assume that

$$supplier = \{Jones, Wilson, Brown, Thomas\}\\ depot = \{Cardiff, London, Sheffield, Manchester\}\\ product = \{nut, bolt, hanger, screw\}$$

(i) $\{(nut, 4), (bolt, 1), (hanger, 7), (screw, 4)\}$
(ii) $\{(Jones, Cardiff), (Wilson, London), (Brown, Sheffield), (Thomas, Manchester)\}$
(iii) $\{(Jones, bolt), (Wilson, bolt), (Wilson, nut), (Brown, hanger), (Thomas, screw)\}$
(iv) $\{(Jones, nut)\}$
(v) $\{(Jones, bolt), (Wilson, hanger)\}$

■

Worked example 7.2 The following is an excerpt from the system specification of a monitoring system for a chemical plant.

> The monitoring system is intended to monitor the operation of five chemical reactors. Every reactor is connected to only one cluster of temperature and pressure sensors, a cluster that is in use is connected only to one reactor; not all the clusters are necessarily in use at the same time.

What type of function can be used to model the operation of the plant?

Solution The fact that a reactor is connected to a cluster can be modelled by an injective function over *reactors* × *clusters*, where *reactors* is the set of reactors in the monitoring system and *clusters* is the set of clusters in the system. If all the clusters were in use, then the function would be total and bijective.

■

Functions can be applied to an element in their domain to yield an element from their range. This is written as the function name followed by the element name. Thus, if *connected* is a function over *computers* × *lines* and has a value

$$\{(VAXA, comm1), (VAXAB, comm2),$$
$$(VAXC, comm3), (VAXD, comm1)\}$$

then the expression

$$connected\ VAXA$$

is equal to *comm*1

Since functions are only a special type of relation the operators which were described in the previous chapter can be used. For example, if the partial injective function *HasAccount* over *customers* × *AccountNos* models the fact that a customer has an account and the partial function *IsLocated* over $AccountNos \times banks$ models the fact that an account is held in a bank, then the function

$$HasAccount \mathbin{;} IsLocated$$

can be used to model the relationship between a customer and his bank. Another example is the expression

$$\#((OverdCustomers \lhd (HasAccount \mathbin{;} IsLocated)) \rhd \{ManchesterA\})$$

which represents the number of overdraft accounts in the *ManchesterA* branch, and where the set *OverdCustomers* is the collection of customers who have overdrafts. The movement of an account of a customer *cust* from a branch to another branch *NewBranch* can be written as

$$IsLocated' = IsLocated \oplus \{(HasAccount\ cust, NewBranch)\}$$

Worked example 7.3 The filing system of a computer is modelled by a function *HasFiles* over $users \times \mathbb{P}\ files$ where if a user has no files then he or she is mapped to the empty set. A subset of the users have special privileges; this subset is modelled by the set *SpecialUsers*. Write down the predicate which states that these special users are allowed to own up to *MaxFiles* files.

Solution The predicate is shown below.

$$\forall u : SpecialUsers \bullet \#HasFiles\ u \leq MaxFiles$$

The predicate uses the universal quantifier to range over all the members of the set *SpecialUsers*.

■

Two functions which will be used frequently in the rest of this book are *pred* and *succ*. The latter is over $\mathbb{N} \times \mathbb{N}$ while the former is over $\mathbb{N}_1 \times \mathbb{N}$. *Pred* returns the predecessor of its argument, while *succ* returns its successor. Thus, the functions are

$$pred = \{(1,0),(2,1),(3,2),(4,3),\ldots\}$$

$$succ = \{(0,1),(1,2),(2,3),(3,4),\ldots\}$$

7.2 Higher-order functions

The functions already described in this chapter have been comparatively simple. Normally, they have involved ranges and domains which contain simple objects such as natural numbers or monitor names. While such functions are adequate for a large proportion of specifications, there is often a need for more complicated functions.

As an example consider the memory allocation subsystem of an operating system which keeps track of the addresses of processes in memory. A process will be characterized by its start address and its finish address. If processes are characterized by name, then they can be modelled by a function which maps process names to $\mathbb{N} \times \mathbb{N}$.

Worked example 7.4 The partial function *MemMap* maps process names p into pairs which represent the start address and finish address of processes. These addresses are expressed in bytes. Write an expression which gives the number of processes that occupy more than 2000 bytes of storage. Assume that *addresses* is the set of all possible addresses in memory.

Solution The expression is

$$\#\{\ p : \mathrm{dom}\ MemMap \mid \exists\ a, b : addresses \bullet (b - a) > 2000 \\ \wedge\ (a, b) = MemMap\ p\ \}$$

It is the cardinality of the set of processes whose pairs differ by more than 2000.

■

Worked example 7.5 Write down a predicate which describes the result of removing all those processes with a length greater than 10 000 bytes from *MemMap*.

Solution The predicate is shown below

$$MemMap' = \{\ p : \mathrm{dom}\ MemMap \mid \exists\ a, b : addresses \bullet \\ (b - a) > 10\,000 \wedge (a, b) = MemMap\ p\ \} \lhd MemMap$$

The predicate relates the value of *MemMap* before removal of the specified processes with its value after removal of the processes. It involves subtracting all those processes with a length greater than 10 000 from *MemMap*.

■

Another example of the use of more complicated functions occurs in the modelling of part of a stock control system which deals with commodities stored at various warehouses. The fact that a series of commodities is stocked at a warehouse can be modelled by means of a partial function *stores* which maps warehouse names to a set of commodity codes. In order to extract the commodities stored at, say, Manchester, all that would be needed to be written would be

$$stores\ Manchester$$

Worked example 7.6 Assume that as part of the stock control system described above a command is required which displays the number of different commodities stored at a location *city*. Write down an expression involving *stores* which gives the required number.

Solution The expression is

$$\#(stores\ city)$$

The expression is the cardinality of the set to which the function *stores* maps *city*.
■

Worked example 7.7 The customer for the stock control system described above requires that a facility be provided which enables new commodity codes for a new city to be added to a particular warehouse. Assume that the user types in the set of new commodity codes *CommCodes* and a new city *CityName*. Write a predicate which expresses the effect of this on *stores*.

Solution The effect of the command is shown below.

$$stores' = stores \cup \{(CityName, CommCodes)\}$$

The command will have the effect of forming the union of *stores* with the singleton set containing the pair $(CityName, CommCodes)$.
■

Another frequent use of complicated functions involves functions where the domain or the type of the range is itself a function. These are known as **higher-order functions**. An example of the use of a higher-order function might occur in a system for keeping track of reservations for a set of rooms in a hotel. The fact that a room is to be occupied by a guest might be modelled by a function which maps room numbers into guest names.

However, if one wishes to keep track of the occupancy of rooms over a time period, then a further function *bookings* is required which maps dates into the previous function. A simple example will make this clear. Let us make the gross simplifying assumptions that the hotel has only three rooms; that booking details are kept for only two days in advance, and that dates are expressed as an integer relative to the beginning of the year. If the value of the function is

$$\{(24, \{(1, Jones), (2, Ince)\}), (25, \{(1, Wilson), (2, Timms), (3, Yeo)\})\}$$

then the function maps day 24 into the function

$$\{(1, Jones), (2, Ince)\}$$

which maps room numbers into occupant's names, and maps the day 25 into the function

$$\{(1, Wilson), (2, Timms), (3, Yeo)\}$$

which again maps room numbers into occupant's names. Each element of the function is a pair. The first element is a date and the second element is a function which maps room numbers into the names of occupants. Thus, if the name of the occupant of *RoomNo* for date *day* is required, then

$$bookings\ day\ RoomNo$$

would be written.

Worked example 7.8 An operation required for the hotel booking system outlined above is that of adding a new day *NewDay* to the function *bookings*. When this day is added the rooms in the hotel will be assumed to be not booked. Write down the predicate which describes this event.

Solution The effect is straightforward:

$$bookings' = bookings \cup \{(NewDay, \varnothing)\}$$

The new value of *bookings* will be the union of the old value together with the pair made up of *NewDay* and the empty set. The latter signifies no bookings for that day.

■

Worked example 7.9 Write down an expression which will deliver the number of rooms which have no bookings at all for the next two days.

Solution The expression is

$$\#\{\ n : \mathbb{N} \mid bookings\ n = \varnothing\ \}$$

This is the cardinality of the set which is formed by specifying those natural numbers which *bookings* map to an empty set.
■

7.3 Modelling a version control system

A version control system is a software tool which keeps track of different versions of a software system. Many large software systems exist in a considerable number of different versions. There are a number of reasons for this: one reason is that a software system may be required to operate in a number of different hardware configurations; each configuration may require a slightly different version of the original system. Another reason is that, as a software system evolves during maintenance, changes will be made to that system; after a time all the changes are normally aggregated and a new version of the system formed and distributed to its users.

The text of part of a specification for a very simple version control system is shown below, and the formal specification of the effect of each command is interspersed with this text. The effect of errors is not included in the specification.

> 1. The purpose of the configuration management system is to keep track of the large number of different versions of software packages produced by International Software Inc.
>
> 2. Each version of a named software system is given a unique integer identifier. The first version has the value 1.
>
> 3. Each version of the software system will contain a number of named program units. Each program unit will exist as a number of different versions. These will be uniquely identified by an integer. The first version of a program unit has the value 1.

The statement of requirements has, so far, indicated that two sets of objects will be required. The first set *SystemNames* will be the set of all possible system names. The second set will be the set of all possible program unit names *ProgramUnitNames*.

A function is also required which maps program unit names into version numbers; this function is used to relate the units making up a system to their version numbers. Also a function *SysVersions* is required which maps *SystemNames* into functions which map system version numbers into the former type of function; this function is used to relate system names to the version numbers of systems and the program units that make up the system. It is assumed that $\mathbb{N}$ will be the set from which program unit version numbers and system version numbers are taken. Thus, the structure of *SysVersions* is

a function mapping elements of *SystemNames* into
  a set of functions each of which map elements of $\mathbb{N}$ into
    a set of functions which map elements of *ProgramUnitNames* into
      the set $\mathbb{N}$

The remaining parts of the text describe the commands and their effects.

4. A version is formed from a previous version in three ways: first, by replacing existing program units in a version with new versions for those program units; second, by removing program units from the version; and third, by adding program units to a version.

5. A number of different commands are to be provided as part of the version control system. They are detailed below.

6. The SETUP command. This command sets up the first version of the system. The user provides the system name together with a list of program units and the version control system stores the program unit names as the first version of the system. The system will assume that the program unit names entered will be the initial versions.

Assuming that the name of the set of program unit names provided is *UnsProvided* and the name of the new system is *sysnm*, then the result of this command can be specified as

$$\begin{aligned} & SysVersions' = SysVersions \cup \\ & \quad \{(sysnm, \{(1, \{\ unit : UnsProvided \bullet (unit, 1)\ \})\})\} \end{aligned}$$

7. The DISPLAY command. This command expects the version number and the name of a system. It will then display the names of the program units which make up that version of the system.

The effect of the command is to display the set

$$\mathrm{dom}(SysVersions\ sysnm\ VersionNo)$$

where *sysnm* is the name of the system and *VersionNo* is the version number.

8. The DIFF command. This expects a version number n of a system together with the system name. It will then display those program units which are not in the nth version of the system but which are in the $(n+1)$th version.

The set produced by this command will be

$$\mathrm{dom}(SysVersions\ sysnm(n+1)) \setminus \mathrm{dom}(SysVersions\ sysnm\ n)$$

Again this assumes that *sysnm* is the name of the system.

> 9. The COMMON command. This command expects a version number n of a system together with the system name. It will then display those program units which are in both the nth version and the $(n+1)$th version.

The set that is displayed is shown below

$$\mathrm{dom}(SysVersions\ sysnm(n+1)) \cap \mathrm{dom}(SysVersions\ sysnm\ n)$$

The common modules are given by the intersection of the expression formed in the previous command specification.

> 10. The NUMIN command. This command expects the version number of a system and a system name. It then displays the number of program units in the system.

The number displayed will be

$$\#(SysVersions\ sysnm\ n)$$

which is the cardinality of the set of program units.

> 11. The NUMVERSIONS command. This command displays the number of current versions of a system. The command expects a system name.

The number is given by

$$\#(SysVersions\ sysnm)$$

which is just the cardinality of the function.

> 12. The FORMVERSION command. This command forms a new version of a software system. The user provides the name of the system together with a set of program unit names and corresponding program unit version numbers. The effect of the command is to form a new version of the system comprising the program units.

If the name of the system is given by *sysnm* and the set of pairs representing program unit names and version numbers is *UnitSet* then the effect of the command can be specified as

$$\begin{array}{l} SysVersions'\ sysnm = SysVersions\ sysnm \cup \\ \{(\#(SysVersions\ sysnm) + 1, UnitSet)\} \end{array}$$

All this does is to form a new set consisting of the units to be added and an integer one bigger than the current largest version number.

13. The OCCUPIES command. This command expects a program unit name and a system name. It will display the number of versions of the software system in which the program unit participates.

The number of versions of the software system which contain the program unit is given by the expression

$$\#\{\ vn : \mathbb{N} \mid UnitName \in \mathrm{dom}(SysVersions\ sysnm\ vn)\ \}$$

where *UnitName* is the name of the program unit provided and *sysnm* is the name of the system.

7.4 Functions as lambda expressions

An alternative way of writing functions which is often used in mathematics is known as a **lambda expression**. The general form of a lambda expression is

$$\lambda\ signature \mid predicate \bullet term$$

It has a similar form to the notation for constructive set specification. The signature establishes the types of the variables used. The predicate gives a condition which each first element of every pair in the function must satisfy; the term gives the form of the second element of each pair in the function. An example of a lambda expression is

$$\lambda\, m : \mathbb{N} \mid m > 4 \bullet m + 5$$

It denotes the infinite function

$$\{(5, 10), (6, 11), \ldots\}$$

Another example is the function

$$\lambda\, x : 0 \,..\, 10 \mid (x, x^2)$$

which is a finite function which maps natural numbers between 0 and 10 to a pair whose first element is the natural number and the second element is its square, i.e.

$$\{(0, (0, 0)), (1, (1, 1)), (2, (2, 4)), \ldots, (10, (10, 100))\}$$

Worked example 7.10 Examine the lambda expression

$$\lambda\, x : \mathbb{N} \mid x < 3 \bullet x^2$$

What is its value?

Solution Its value is

$$\{(0,0),(1,1),(2,4)\}$$

The predicate constrains it to the first three natural numbers.

■

In general, a lambda expression maps types made up from variables of the signature into the expression represented by the term for which the predicate is true. Thus,

$$\lambda\, a, b, c : \mathbb{N} \mid a + b = c \bullet a^2 + b^2 + c^2$$

specifies the function

$$\{((0,0,0),0),((0,1,1),2),((1,0,1),2),((1,1,2),6),\ldots\}$$

since each (a, b, c) in every pair satisfies the predicate $a + b = c$ and the second element of each pair is $a^2 + b^2 + c^2$.

Exercise 7.4
If *files* is the set

$$\{upd, text, ed1, ed2, ed3, tax1, tax2\}$$

and *users* is the set

$$\{Timms, Jones, Wilkins, Wilson\}$$

and *size* is a partial function over *files* $\times\ \mathbb{N}$ with the value

$$\{(upd,45),(text,175),(ed1,105),(ed2,95)\}$$

then write down the values of the following expressions.

(i) $\lambda\, x : 3\,..\,15 \mid x^2 \leq 9 \bullet x$
(ii) $2\,..\,4 \lhd (\lambda\, x : \mathbb{N} \mid x < 10 \bullet x)$
(iii) $\lambda\, x : \mathbb{N} \mid x < 10 \bullet x^2$
(iv) $\lambda\, x : \mathrm{ran}\ size \mid x > 53 \bullet x^2 + 10$
(v) $\lambda\, x : \mathbb{N} \mid x = \mathrm{ran}\ size$
(vi) $\{2,3\} \lhd \lambda\, x : \mathbb{N} \bullet x^2$
(vii) $size \mathbin{;} (\lambda\, x : \mathbb{N} \mid x <> 100 \bullet x + 10)$
(viii) $(\lambda\, x : \mathbb{N} \mid x <> 5 \land x < 10 \bullet x) \cup (\lambda\, x : \mathbb{N} \mid x^2 = 16 \bullet x^3)$

■

7.4.1 Curried functions

One of the most powerful facilities that lambda abstraction allows the specifier is known as **currying**. As an example of this consider the function

$$\lambda\, x : \mathbb{N} \bullet \lambda\, y : \mathbb{N} \bullet x + y$$

This represents the infinite function which forms the sum of two natural numbers. This function can be regarded as a function which maps natural numbers into a function which itself maps natural numbers into natural numbers. Not only does this allow the specifier to write

$$(\lambda\, x : \mathbb{N} \bullet \lambda\, y : \mathbb{N} \bullet x + y) 8\ 9$$

which delivers 17, but also allows the specifier to generate a family of functions. For example, the function that adds 3 to a number can be written as

$$(\lambda\, x : \mathbb{N} \bullet \lambda\, y : \mathbb{N} \bullet x + y) 3$$

This generates the function

$$\{(0,3),(1,4),(2,5),\ldots\}$$

and the function that adds 2098 to a number can be written as

$$(\lambda\, x : \mathbb{N} \bullet \lambda\, y : \mathbb{N} \bullet x + y) 2098$$

The treatment of functions with two or more arguments is named after the logician H. B. Curry.

7.5 Sequences as functions

In many system specifications there is a need to model the fact that a series of objects is to be associated with natural numbers and an implied ordering. For example the excerpts from the following specifications all indicate some ordering is inherent in the relations that are being described.

> ... Each message to be printed at the remote printer will be queued according to the priority of the message.
>
> ... There will be 200 consecutive memory slots. Each program currently awaiting execution will occupy adjacent slots. The programs will be stored in ascending order of size.
>
> ... On receipt of the LISTSUM command the program will print out stock identities in descending order of stock holding.

Because such situations are very common in specifications they can be modelled using a special function known as a **sequence**. A sequence of type T is a partial function from $\mathbb{N}_1$ to T where the domain of the function is of the form $1 \mathinner{..} n$. For example, if *files* is the set

$$\{upd, tax1, tax2, edtax, newupd, oldupd, credit1, credit2\}$$

then all the following are examples of sequences of files

$$\{(1, upd)\}$$
$$\{(1, tax1), (2, upd)\}$$
$$\{(1, tax1), (2, upd), (3, oldupd), (4, credit1)\}$$
$$\{(1, tax1), (2, upd), (3, oldupd), (4, credit1), (5, edtax), (6, credit1)\}$$

In this book we shall write sequences surrounded by angle brackets. For example, the last sequence above would be written as

$$\langle tax1, upd, oldupd, credit1, edtax, credit1\rangle$$

Formally, a sequence is a function which maps members of $\mathbb{N}$ to members of a set T and is defined as

$$\mathrm{dom}\, S = 1 \mathinner{..} \#S \wedge \mathrm{ran}\, S \subseteq T$$

i.e. a sequence of objects of type T is a function from natural numbers to T such that the domain of the function is the set of numbers from 1 to the cardinality of the function and the range is taken from T. This definition ensures that the sequence will contain the consecutive numbers from 1 to $\#S$ in the domain of the function.

Given this definition it is relatively straightforward to extract an element of a sequence. For example, if *PrintQueue* is a sequence of files then *PrintQueue* 1 gives the first element of the sequence while

$$PrintQueue\ \#PrintQueue$$

gives the final element of the sequence.

A number of operators can be defined which operate on sequences. The first is the *head* operator which is a function that returns the first object in a sequence. This can be defined using the lambda form as

$$\lambda\, s : S \mid s \neq \varnothing \bullet s\ 1$$

where s is a sequence of objects of an arbitrary type. The operator *front* returns the first $n - 1$ elements of a sequence of length n. It is defined as

$$\lambda\, s : S \mid s \neq \varnothing \bullet (1 \mathinner{..} \#s - 1) \lhd s$$

Worked example 7.11 The *tail* operator extracts from a sequence of length n the last $n - 1$ elements. Use the *succ* function to define *tail*.

Solution The *tail* operator is defined as

$$\lambda\, s : S \mid s \neq \varnothing \bullet \{0\} \lhd (succ \,{}_{9}^{\circ}\, s)$$

The *succ* function is composed with s in order to subtract one from the first element of each pair. This gives a function which is identical to s except that the first element of each pair has been decreased by one. The first element of this sequence can then be removed using the domain subtraction operator.

∎

Finally, the function *last* can be defined. *last* gives the last element of a sequence s and is defined as

$$\lambda\, s : S \mid s \neq \varnothing \bullet s\ \#s$$

Worked example 7.12 A system specification for an operating system requires that the computer operator be able to remove the nth to mth entries in a queue of programs awaiting execution. Define a function using lambda notation which, when applied to a sequence of programs, gives the set of nth to mth programs.

Solution The definition is

$$\lambda\, n, m : \mathbb{N};\ s : S \mid n > 0 \land m \geq n \land m \leq \#s \bullet \mathrm{ran}(n \mathrel{..} m \lhd s)$$

where s is a sequence of programs. The predicate

$$n > 0 \land m \geq n \land m \leq \#s$$

specifies the condition under which the function is applied since if n and m are out of range or $n > m$ then the function is undefined.

∎

The functions defined above have all involved retrieving items from sequences. There is also a need for sequences to be constructed from other sequences. This is the function of the **concatenation** operator $\frown$. It takes two sequences and places the second sequence after the first. Thus, if *first* is the sequence

$$\langle file1, file7 \rangle$$

and *second* is the sequence

$$\langle newfile, oldupd, oldtax, newed \rangle$$

then

$$first \frown second = \langle file1, file7, newfile, oldupd, oldtax, newed \rangle$$

The definition of $\frown$ is

$$\langle a_1, a_2, \ldots, a_n \rangle \frown \langle b_1, b_2, \ldots, b_m \rangle = \langle a_1, a_2, \ldots, a_n, b_1, b_2, \ldots, b_m \rangle$$

$$\langle \rangle \frown A = A \frown \langle \rangle = A$$

The first line defines concatenation for two non-empty sequences and the second line defines concatenation where one of the sequences is empty.

Worked example 7.13 Two queues containing files are maintained by a computer's operating system. The queues are named *sys*1 and *sys*2. An operation is required which will concatenate the two queues and form a new queue *nqueue*. If *sys*1 is longer than or equal to *sys*2 then it is placed at the front. However, if *sys*2 is longer than *sys*1 then the former is placed at the front. Write down a predicate which describes this. (Hint: since the sequences are really just sets, operators such as # can be used.)

Solution The predicate is

$$\#sys1 \geq sys2 \Rightarrow nqueue = sys1 \frown sys2 \wedge$$
$$\#sys1 < sys2 \Rightarrow nqueue = sys2 \frown sys1$$

The two alternatives are implemented by means of the $\Rightarrow$ operator.
■

7.6 Applying sequences—a print spooler

A computer operating system normally has more requests for printing facilities than printing facilities available. In order to cope with this the operating system of the computer will ensure that when a program produces any output it is first copied to a file. This file joins all other files awaiting printing in a queue or set of queues. The following specification describes the operator facilities available for manipulating such queues.

Spool queues—operator facilities

1. A number of commands will be available for the computer operator which enable him or her to manipulate queues or items in a queue.

2. Each output peripheral connected to the computer will be associated with a queue.

3. The SIZE command will return the number of entries in a queue.

4. The EMPTYSIZE command will return the number of queues which are empty together with the name of the peripheral associated with a particular queue.

5. The REMOVE command will remove a specified file from the queue associated with a particular peripheral.

6. The ADD command will remove all the items from a queue associated with a peripheral and add them to the end of another queue.

7. The MOVE command will remove an item from a queue associated with a peripheral and place it at the end of another queue.

8. The TOTALQUEUE command will return with the number of queues currently active, i.e. those queues which contain at least one entry.

Operators of the computer system will normally specify a peripheral name when using the above commands. For example, the operator will be more interested in the size of a queue associated with a peripheral than the number of items in a named queue. Thus, it seems natural to model the spooling system by means of a partial function from peripheral names to sequences of spool files. This can be done because each peripheral is associated with only one queue. If this were not the case, then a relation would have to be used. Therefore, let us assume the existence of a partial function *PerAc* which maps *PeripheralNames* to a sequence of files. Each command can be specified in terms of operations on this function. In the specification that follows it is assumed that the operator has typed in a valid command and we are not concerned with error processing.

The SIZE command assumes that the operator has typed in *PeriphName* as a peripheral and the system returns the value of the queue size in *size*. The effect of the command can be specified in terms of the predicate

$$\begin{array}{l} PeriphName \in \operatorname{dom} PerAc \wedge size = \#(PerAc\ PeriphName) \wedge \\ PerAc' = PerAc \end{array}$$

The first conjunct states that the peripheral has a queue associated with it, the second conjunct establishes the value of *size*, and, lastly, the third conjunct specifies the fact that the command does not affect *PerAc*.

The EMPTYSIZE command forms a set of peripheral names *PeriphSet* such that the function *PerAc* maps each element of the set to an empty sequence. The number of queues currently empty *NumberEmpty* will be equal to the cardinality of this set. The predicate expressing this is

$$PeriphSet = \{ per : peripherals \mid PerAc\ per = \varnothing \} \land$$
$$NumberEmpty = \#PeriphSet \land$$
$$PerAc' = PerAc$$

Worked example 7.14 Specify the REMOVE command. Assume that the operator has typed in a file name *FileName* and a peripheral name *PeriphName*.

Solution The REMOVE command can be specified by using the domain restriction operator

$$PeriphName \in \operatorname{dom} PerAc \land$$
$$FileName \in \operatorname{ran}(PerAc\ PeriphName) \land$$
$$QueuePosn = (PerAc\ PeriphName)^{-1}\ Filename \land$$
$$PerAc'\ PeriphName = (1\ ..\ (QueuePosn - 1) \lhd PerAc\ PeriphName$$
$$\cup succ \mathbin{;} ((QueuePosn + 1\ ..\ \#(PerAc\ PeriphName))$$
$$\lhd PerAc\ PeriphName)$$

The first conjunct specifies that the peripheral name must have a queue associated with it. The second conjunct states that *FileName* should be in the queue associated with *PeriphName*. The third conjunct gives the position in the queue of *FileName*. This is expressed using the inverse of the sequence which represents the queue of files. This, of course, assumes that the function is bijective. If it were not, then the command would not be meaningful as there would be more than one instance of a file name in a particular queue. The final conjunct defines the new sequence. It is the old sequence minus the extracted item.
■

The ADD queue command will remove those elements from a queue associated with a peripheral, say *per*1, and add them to another queue associated with a peripheral *per*2. The specification is

$$per1 \in \operatorname{dom} PerAc \land$$
$$per2 \in \operatorname{dom} PerAc \land$$
$$PerAc'\ per2 = (PerAc\ per2) \frown (PerAc\ per1) \land$$
$$PerAc'\ per1 = \varnothing$$

The first two conjuncts state that both peripherals must have queues associated with them. The third conjunct states that the *per*2 queue becomes the old *per*2 queue with the *per*1 queue joined to it. The final conjunct states that the *per*1 queue becomes empty.

The MOVE command will remove an item *FileName* from the queue associated with *per*1 and place it at the end of the queue associated with *per*2. The specification for this command is

$$
\begin{aligned}
& per1 \in \mathrm{dom}\, PerAc \wedge \\
& per2 \in \mathrm{dom}\, PerAc \wedge \\
& Filename \in \mathrm{ran}(PerAc\ per1) \wedge \\
& QueuePosn = (PerAc\ per1)^{-1} FileName \wedge \\
& PerAc'\ per1 = (1 \mathinner{..} (QueuePosn - 1)) \lhd PerAc\ per1 \cup \\
& \qquad succ \mathbin{;} ((QueuePosn + 1 \mathinner{..} \#(PerAc\ per1)) \lhd PerAc\ per1) \wedge \\
& PerAc'\ per2 = PerAc\ per2 \cup \{(\#(PerAc\ per2) + 1, Filename\}
\end{aligned}
$$

Worked example 7.15 Specify the TOTALQUEUE command.

Solution This command can be specified by calculating the cardinality of the set of all peripherals which are mapped by *PerAc* to non-empty queues. If *TotQueues* is the total number of non-empty queues, then this command can be specified as the predicate

$$
\begin{aligned}
& TotQueues = \#\{\ per : peripherals \mid PerAc\ per = \langle\rangle\ \} \wedge \\
& PerAc' = PerAc
\end{aligned}
$$

The second conjunct specifies that the command does not affect *PerAc*
■

8

The specification language Z

Aims

- To describe the schema, the main structuring device in the Z specification language.
- To describe how schemas can be included within each other.
- To outline how software systems can be described in the Z specification language.

Although in the previous chapters computer systems have been described in precise terms, there has been little stress on the syntax of the mathematical notations used. This part of the book remedies this state of affairs. It describes a language called Z which is based on typed set theory and which allows a concise expression of the functional properties of computer systems. One of the principal features of Z is an object known as a **schema**.

8.1 Schemas

A schema is a structuring device that allows us to specify the stored data of a system and the operations that access that stored data. A schema is written in two forms. The first form is partly graphical and is shown below

$$\begin{array}{|l} a : \mathbb{N} \\ b : \mathbb{P}\ \mathbb{N} \\ \hline a \in b \end{array}$$

The box contains two sections, the part above the middle line is known as the **signature** and the part below the middle line is known as the **predicate**. The signature introduces variables and assigns them a set theoretic type. This is similar to declarations in a programming language.

The schema above introduces two variables a and b and states that a will be a natural number and b will be a set of natural numbers. The only difference between the programming language declarations and those found in a schema is that since a computer has a finite store the range of the values of the former is limited.

The predicate of a schema refers to the variables introduced in the schema or global variables in other schemas and relates the values of these variables to each other. For example, the predicate in the schema above asserts that a is contained in b.

An alternative way of expressing schemas is to use a linear form. Here a schema is constructed by listing the signature followed by the predicate with | marking their separation and with the whole box enclosed in square brackets. Thus an alternative to the box structure above would be

$$[a : \mathbb{N};\ b : \mathbb{P}\ \mathbb{N} \mid a \in b]$$

Whenever a series of predicates is written in a schema with each one on a separate line, then the total predicate represented by that schema is the conjunction of the predicates on each line. For example, the schema

$$\begin{array}{|l}\hline a, b : \mathbb{N} \\ c : \mathbb{P}\ \mathbb{N} \\ \hline a \in c \\ b \in c \\ \hline \end{array}$$

is equivalent to

$$\begin{array}{|l}\hline a, b : \mathbb{N} \\ c : \mathbb{P}\ \mathbb{N} \\ \hline a \in c \wedge b \in c \\ \hline \end{array}$$

or

$$[a, b : \mathbb{N};\ c : \mathbb{P}\ \mathbb{N} \mid a \in c \wedge b \in c]$$

Schemas often need to be named because they will be referred to in other schemas. Naming is achieved by labelling a schema with a name, for example,

$$\begin{array}{|l}\hline \textit{MonCondition} \\ \hline \textit{MonNo} : \mathbb{N} \\ \textit{AvailableMonitors} : \mathbb{P}\ \mathbb{N} \\ \hline \textit{MonNo} \in \textit{AvailableMonitors} \\ \hline \end{array}$$

The equality symbol $\widehat{=}$ is also used to label the linear form of a schema; for example,

$$MonCondition \ \widehat{=}$$
$$[MonNo : \mathbb{N};\ AvailableMonitors : \mathbb{P}\ \mathbb{N} \mid MonNo \in AvailableMonitors]$$

is equivalent to the above schema. The signature in a schema can introduce variables of any set theoretic type. They range from natural numbers up to complicated higher-order functions. Whenever functions and relations are defined in a signature, their type is designated by the types of their domains and range, together with a symbol which gives the type of the function or relation. A full list of symbols used to distinguish the types of function follows.

a relation	$\leftrightarrow$
a partial function	$\nrightarrow$
a total function	$\rightarrow$
a partial injection	$\rightarrowtail\!\!\!\!\!\!+$
a total injection	$\rightarrowtail$
a partial surjection	$\twoheadrightarrow\!\!\!\!\!\!+$
a total surjection	$\twoheadrightarrow$
a bijection	$\rightarrowtail\!\!\!\!\!\twoheadrightarrow$

Some examples of schemas using such signatures are

accounts

$HoldsAccount : AccountNo \nrightarrow customer$

$HoldsAccount \neq \varnothing$

upd

$CustomerNo : \mathbb{N}$

$UpGroup : (\mathbb{N} \rightarrow \mathbb{N}) \rightarrow (\mathbb{N} \rightarrow \mathbb{N})$

$\#\,\mathrm{dom}\,\mathrm{dom}\ UpGroup > 50$

SqState

$SquareAdd : \mathbb{N} \rightarrow \mathbb{N}$

$\forall n : \mathbb{N} \bullet SquareAdd\ n = n^2 + 12$

The first schema *accounts* contains a partial function *HoldsAccount* in its signature. Its predicate asserts that there will be at least one element in the function. The second named schema *upd* introduces two variables *CustomerNo* and *UpGroup*. The former is a natural number while the latter is higher-order function which takes its domain from functions which map natural numbers into natural numbers and its range from functions which map natural numbers to natural numbers. The predicate asserts that the cardinality of the domain of the functions which are in the domain of *Upgroup* is greater than 50. Lastly, the schema *SqState* defines a total function *SquareAdd* which is

$$\{(0,12),(1,13),(2,16),(3,21),\ldots\}$$

Worked example 8.1 Write down a schema *BlockInv* which introduces two variables. The first variable is a partial function *BlockOwn* which maps natural numbers into sets of natural numbers. The second variable is a natural number *MaxBlocks*. The schema should include a predicate which states that *BlockOwn* will never map a natural number into a set whose cardinality is greater than *MaxBlocks*.

Solution The schema is

```
── BlockInv ──────────────────────────────
│ BlockOwn : ℕ ⇸ ℙ ℕ
│ MaxBlocks : ℕ
├──────────────
│ ∀ n : dom BlockOwn • #(BlockOwn n) ≤ MaxBlocks
└──────────────────────────────────────────
```

■

Worked example 8.2 A fragment of the natural language specification for a filing system follows.

> There will never be more than 30 users of the file system who currently own files, no user is allowed to own more than 25 files, and no file is allowed to be bigger than 500 blocks.

In a Z specification the file store of the system is modelled by a function *stores* which relates user names to sets of file names, and a function *occupies* which relates a file name to the set of natural numbers which represent the blocks that the files occupy. Write down a schema which is equivalent to this description.

Solution The schema is shown below.

$$\begin{array}{|l}
\textit{ExSchem} \\
\hline
stores : UserNames \twoheadrightarrow \mathbb{P}\ Filenames \\
occupies : FileNames \twoheadrightarrow \mathbb{P}\ \mathbb{N} \\
\hline
\#\,\mathrm{dom}\, stores \leq 30 \\
\forall\, name : \mathrm{dom}\, stores \bullet \#(stores\ name) \leq 25 \\
\forall\, file : \mathrm{dom}\, occupies \bullet \#(occupies\ file) \leq 500 \\
\hline
\end{array}$$

■

8.2 Supporting facilities in Z

As well as schemas Z features a number of facilities which enable system specifications to be built up easily. The first facility allows the specifier to declare the sets which are basic types in a specification: those sets which are assumed to exist, that can be used in specifications, and do not require any further definition. These sets are declared by enclosing them in square brackets. For example, the declaration

$$[CARS, SPAREPARTS]$$

declares two sets used in a spare parts warehouse application. This book will adopt the convention that such sets are written using capital letters.

Another facility allows the specifier to declare a set with a small number of members. In the declaration the members are separated by the | symbol. For example,

$$cars ::= Rover \mid Ford \mid GeneralMotors \mid Fiat$$

$$owners ::= Thomas \mid Ince \mid Roberts \mid Williams \mid Jewson$$

defines the set *cars* to have four members and the set *owners* to have five members.

The symbol == introduces a global constant. Thus,

$$MaxIds == 50$$

declares the constant *MaxIds* as having the value 50 with *MaxIds* being able to be used throughout the specification in which it was declared.

The final facility is called an **axiomatic description**. It declares a global variable which has some constant associated with it. An axiomatic description is written in a similar way to a schema. However, it lacks the top and bottom lines. Examples of axiomatic descriptions are shown below.

$$\begin{array}{|l} MemberValue : \mathbb{N} \\ \hline MemberValue < 100 \end{array}$$

This declares a global *MemberValue* which is a natural number less than 100. Another example is shown below of a function *TopHalf* which maps natural numbers into sets of natural numbers. The constraint specified in the predicate is that the range of the function will never contain sets which have more than 20 members

$$\begin{array}{|l} TopHalf : \mathbb{N} \rightarrow \mathbb{P}\ \mathbb{N} \\ \hline \forall\, n : \mathrm{ran}\ TopHalf \bullet \#n \leq 20 \end{array}$$

Such axiomatic descriptions can refer to other global variables. An example is shown below of the axiomatic description of a set *NewUsers*

$$\begin{array}{|l} NewUsers : \mathbb{P}\ users \\ \hline NewUsers = \{\ u : users \mid u \in WeekUsers \wedge life\ u < 10\ \} \end{array}$$

This describes the set whose members are taken from the set *WeekUsers* which, when the global function *life* is applied, yields a value less than 10. In this axiomatic description two globals *WeekUsers* and *life* are referenced, the first being a set of users, the second being a function which maps users to natural numbers.

8.3 The structure of Z specifications

One of the major features of Z is that it provides a framework within which a specification can be developed and presented incrementally. A full Z specification can be constructed from other schemas which themselves can be expressed in terms of other schemas etc. Each subsidiary schema can be referred to by schemas which use it.

This is similar to the presentation of a software design or a program. A well-written program will consist of a series of program units which can be read and understood in isolation. A Z specification will consist of schemas which can be developed and presented incrementally with individual schemas being able to be understood in isolation.

8.3.1 Schema inclusion

The means by which schemas can be referred to by other schemas is known as **schema inclusion**. As an example consider the schema *SetInv*

$$
\begin{array}{|l}
\textit{SetInv} \\
\hline
upper, lower : \mathbb{P}\ \mathbb{N} \\
MaxSize : \mathbb{N} \\
\hline
\#upper + \#lower \leq MaxSize \\
\hline
\end{array}
$$

which introduces two sets of natural numbers *upper* and *lower*, and states that the sum of their cardinalities will never exceed *MaxSize*. If this schema is required by another schema, then it is written in the signature part of the schema that requires it. For example, the schema *MidInv* refers to *upper* and *lower* and hence uses *SetInv*:

$$
\begin{array}{|l}
\textit{MidInv} \\
\hline
middle : \mathbb{P}\ \mathbb{N} \\
SetInv \\
\hline
middle \subset upper \cup lower \\
\hline
\end{array}
$$

Since it uses *upper* and *lower* from *SetInv*, the schema is included in *MidInv*. The effect of including one schema in the signatures of another schema is to form the union of their signatures and to conjoin their predicates. Thus, *MidInv* is equivalent to

$$
\begin{array}{|l}
\textit{MidInv} \\
\hline
middle : \mathbb{P}\ \mathbb{N} \\
upper, lower : \mathbb{P}\ \mathbb{N} \\
MaxSize : Nat \\
\hline
middle \subset upper \cup lower \\
\#upper + \#lower \leq MaxSize \\
\hline
\end{array}
$$

8.3.2 Events and observations in Z

Z specifications describe three types of entities: states, observations, and events. A **state** is the mathematical structure which models a system. An **event** is an occurrence which is of interest to the specifier. For example, in the specification of a queue subsystem typical events would be: the addition of an entry to a queue; the deletion of the queue; or the calculation of queue length. In the specification of a filing system typical events would be: the creation of a file; the deletion of a file; or the reading of a file. An **observation** is a set theoretic variable whose value can be examined before or after an event has occurred. Typical examples of observations are: a function modelling a file store; a relation modelling the fact that a system user owns a file; and a set of books in an automated library system. States are made up of observations.

There are two types of properties of a state that a Z specification is intended to reflect. The first is the static properties. These are predicates which always hold over the course of time no matter what event occurs. These are often known as **invariants**. The second type of property are those which characterize the effect of an event. These properties are embodied in **observations**.

An example will make this clearer. Suppose that it was necessary to specify a system which kept track of students who have handed in homework assignments. This can be characterized by a state consisting of three sets: the set of all students; the set of students who have handed in assignments; and the set of students who have not handed in assignments.

There are two static properties. The first is that the union of the set of students who have handed in an assignment and the set of students who have not handed in an assignment is equal to the set of students in the class. The second static property is that, since a student cannot simultaneously have handed in an assignment and not have handed in an assignment, then the intersection of the set of students who have handed in an assignment and the set of those who haven't is the empty set.

These properties can be characterized by the schema $\Delta ClassHomework$.

$$
\begin{array}{|l}
\hline
\Delta ClassHomework \\
\hline
class, HandedIn, NotHandedIn : \mathbb{P}\ STUDENTS \\
class', HandedIn', NotHandedIn' : \mathbb{P}\ STUDENTS \\
\hline
HandedIn \cup NotHandedIn = class \\
HandedIn' \cup NotHandedIn' = class' \\
HandedIn \cap NotHandedIn = \varnothing \\
HandedIn' \cap NotHandedIn' = \varnothing \\
\hline
\end{array}
$$

As will be explained in Chapter 10 any schema which represents an invariant property starts with the Δ symbol. The schema above contains a state made up of the components *class*, *HandedIn*, *NotHandedIn* which are sets of students. The observations which make up the state are assigned a set theoretic type $\mathbb{P}\ STUDENTS$ and the predicates characterize the properties of the observations which must always be true. Primes are used, as in the earlier part of the book, to indicate the value of the observations after an operation occurs. For example, the schema states that before any event occurs the union of *HandedIn* and *NotHandedIn* will be equal to *class*

$$HandedIn \cup NotHandedIn = class$$

while after any event has occurred the union of $HandedIn'$ and $NotHandedIn'$ will be equal to $class'$

$$HandedIn' \cup NotHandedIn' = class'$$

The dynamic properties of the system are again characterized by: the names of observations made before an event occurs; the names of those observations which can be made after the event; and a predicate which relates them. For example, in the student assignment specification assume that there are two events. The first is that of a student *stud?* handing in an assignment. The second event is that of the number of students who have handed in an assignment being queried.

```
__HandIn_______________________________
| stud? : STUDENTS
| ΔClassHomework
|______________
| stud? ∈ NotHandedIn
|
| NotHandedIn' = NotHandedIn \ {stud?}
|
| HandedIn' = HandedIn ∪ {stud?}
|
| class' = class
|_______________________________________
```

The signature contains the observations that can be made before and after the *HandIn* event. Primed observations are those made after the event. The inclusion of the schema $\Delta ClassHomework$ adds the invariant properties of the student assignment system. The first predicate states that the student must not already have handed in an assignment. The second and third predicates describe the fact that the student who has handed in an assignment is removed from the *NotHandedIn* set and placed in the *HandedIn* set. The final predicate asserts that the membership of the class is unaffected by the *HandIn* event.

The next event is that of finding the number of students who have handed in an assignment. In order to write down this schema I will introduce a subsidiary schema $\Xi ClassHomework$. This describes the fact that the data in the system will be unaffected

```
__ΞClassHomework_______________________
| class, HandedIn, NotHandedIn : ℙ STUDENTS
| class', HandedIn', NotHandedIn' : ℙ STUDENTS
|______________
| class' = class
|
| HandedIn' = HandedIn
|
| NotHandedIn' = NotHandedIn
|_______________________________________
```

This schema is prefixed by the symbol Ξ, which is a Z convention that states that the schema will not affect stored data.

The schema *AssignQuery* can now be written using the linear schema form as

$$AssignQuery \mathrel{\widehat=} \\ [NotHandedin! : \mathbb{N};\ \Xi ClassHomework \mid NotHandedIn! = \# HandedIn]$$

There is no reason, of course, why the box form of the schema cannot be used. The number of students who have handed in an assignment is placed in the variable *NotHandedIn!*. Another Z convention is used here: output variables are postfixed with an exclamation mark.

Worked example 8.3 A filing subsystem consists of a series of files which are owned by users of an operating system. Each file occupies a series of blocks of storage. The filing system can be modelled by means of a partial function *owns* ($users \rightarrow\!\!\!\!\!+ \mathbb{P}\ FileNames$), the partial function *occupies* ($FileNames \rightarrow\!\!\!\!\!+ \mathbb{P}\ BlockNos$), the set *users* which contains all possible user names, the set *SystemUsers* which is a subset of *users*, the set *BlockNos* which is the set of all possible block numbers, the set *FileNames* which contains all possible file names, and the set *FreeBlocks* which is a subset of *BlockNos*. *Owns* models the fact that a user currently owns a series of files; *occupies* models the fact that a file occupies a series of blocks on a file storage device; and *SystemUsers* are those users who are currently allowed to use the system. No more than *NoUsers* are allowed to use the system. Write down the Z specification for the file system which characterizes its static properties. Make the artificial assumption that each file name created by a user is unique.

Solution The schema is shown below.

$$
\begin{array}{|l}
\textit{FileSystem} \\
\hline
owns : users \rightarrow\!\!\!\!\!+ \mathbb{P}\ FileNames \\
occupies : FileNames \rightarrow\!\!\!\!\!+ \mathbb{P}\ BlockNos \\
SystemUsers : \mathbb{P}\ users \\
FreeBlocks : \mathbb{P}\ BlockNos \\
NoUsers : \mathbb{N} \\
\hline
\# SystemUsers \leq NoUsers \\
\forall file : \mathrm{dom}\, occupies \bullet \exists\, us : \mathrm{dom}\, owns \bullet file \in (owns\ us) \\
\forall file : \mathrm{dom}\, occupies;\ block : BlockNos \bullet \\
\quad block \in (occupies\ file) \Rightarrow block \notin FreeBlocks \\
\mathrm{dom}\, owns = SystemUsers \\
\forall fs1, fs2 : \mathrm{ran}\, owns \bullet fs1 \neq fs2 \Rightarrow fs1 \cap fs2 = \varnothing \\
\hline
\end{array}
$$

It corresponds to the natural language description of its static properties:

> The number of users must not exceed *NoUsers*; if a file is known to the system as occupying a series of blocks, then it is owned by a user; all blocks that make up a file are not free blocks; only system users are allowed to own files; and, finally, no two files are allowed to share blocks.

If the sets *users*, *FilenNames*, and *BlockNos* were basic sets then the Z convention would be that they would be written using capital letters.
■

Given the schema above, a number of events can be defined in terms of their effect on the filing subsystem. These events correspond to commands employed by users. For example, an event which, for a given user *usname*?, removes one of his files *fname*? can be defined as

Remove

$FileSystem$
$FileSystem'$
$usname? : users$
$fname? : Filenames$

$usname? \in SystemUsers$

$fname \in owns\ usname?$

$SystemUsers' = SystemUsers$

$FreeBlocks' = FreeBlocks \cup occupies\ fname?$

$occupies' = \{fname?\} \lhd occupies$

$owns' = owns \oplus \{(usname?, owns\ usname? \setminus \{fname?\})\}$

$NoUsers' = NoUsers$

The first predicate states that the user who employs the command must be a system user. The second predicate states that the name of the file to be deleted must be owned by the user whose name is *usname*? The third predicate states that the set of system users is unchanged by the event. The fourth predicate states that *FreeBlocks* is updated by adding to it those blocks which were contained in the file to be removed. The fifth predicate states that the file name is removed from the *occupies* function since it no longer occupies any blocks. The sixth predicate removes the file from the *owns* function since the user no longer owns the file. The final predicate states that the maximum number of users is unchanged by the event.

FileSystem' is defined as

$$
\begin{array}{|l}
\hline
\textit{FileSystem}' \\
\hline
owns' : users \twoheadrightarrow \mathbb{P}\ Filenames \\
occupies' : FileNames \twoheadrightarrow \mathbb{P}\ BlockNos \\
SystemUsers' : \mathbb{P}\ users \\
FreeBlocks' : \mathbb{P}\ BlockNos \\
NoUsers' : \mathbb{N} \\
\hline
\#SystemUsers' \leq NoUsers' \\
\forall file : \mathrm{dom}\ occupies' \bullet \exists\ us : \mathrm{dom}\ owns' \bullet file \in (owns'\ us) \\
\forall file : \mathrm{dom}\ occupies';\ block : BlockNos \bullet \\
\quad block \in (occupies'\ file) \Rightarrow block \notin FreeBlocks' \\
\mathrm{dom}\ owns' = SystemUsers' \\
\forall fs1, fs2 : \mathrm{ran}\ owns' \bullet fs1 \neq fs2 \Rightarrow fs1 \cap fs2 = \varnothing \\
\hline
\end{array}
$$

The addition of a file *fname?* by a user *usname?* can be defined by the schema *AddUser*

$$
\begin{array}{|l}
\hline
\textit{AddUser} \\
\hline
FileSystem \\
FileSystem' \\
usname? : users \\
fname? : Filenames \\
\hline
fname? \notin (owns\ usname?) \\
usname? \in SystemUsers \\
SystemUsers' = SystemUsers \\
FreeBlocks' = FreeBlocks \\
(owns'\ usname?) = (owns\ usname?) \cup \{fname?\} \\
occupies' = occupies \cup \{(fname?, \varnothing)\} \\
NoUsers' = NoUsers \\
\hline
\end{array}
$$

Worked example 8.4 Write down a schema which describes the effect of removing a user *usname?* and all the files owned by that user from the file store. The schema should have five predicates. The first predicate should assert that the user is one of the current system users. The second predicate should assert that the user has been removed from the set of allowable users. The third predicate should assert that after the *RemoveUser* event has occurred all the files owned by *usname*? have been removed. The fourth predicate asserts that after the event has occurred all the blocks in the files owned by *usname*? have been removed. Finally, the fifth predicate should assert that after the *RemoveUser* event the free blocks occupied by those files owned by *usname*? are returned to the store of free blocks.

Solution The schema is shown below.

$$
\begin{array}{l}
\textit{RemoveUser} \\
\hline
\textit{FileSystem} \\
\textit{FileSystem}' \\
\textit{usname?} : \textit{users} \\
\hline
\textit{usname?} \in \textit{SystemUsers} \\
\textit{SystemUsers}' = \textit{SystemUsers} \setminus \{\textit{usname?}\} \\
\textit{owns}' = \{\textit{usname?}\} \lhd \textit{owns} \\
\textit{occupies}' = \textit{owns usname?} \lhd \textit{occupies} \\
\textit{FreeBlocks}' = \textit{FreeBlocks} \cup \\
\{\ b : \textit{BlockNos} \mid \exists f : \textit{FileNames} \bullet \\
f \in (\textit{owns usname?}) \land b \in (\textit{occupies } f)\ \} \\
\hline
\end{array}
$$

■

9

Operations and objects in Z

Aims

- To describe a large part of the Z toolkit.
- To describe the use of schemas in the definition of generics.
- To provide further practice in the use of schemas.

9.1 The Z toolkit

The primary aim of this chapter is to look at the facilities offered within the Z language. These facilities which include basic data types such as the natural numbers also include operations on set, function and sequence data types. The facilities are frequently referred to as **the Z toolkit**.

9.2 Numbers and sets of numbers

Z recognizes three built-in sets: the natural numbers $\mathbb{N}$, the integers $\mathbb{Z}$: whole numbers which range from minus infinity to plus infinity, and the natural numbers excluding 0. The latter are defined below

$$\mathbb{N}_1 : \mathbb{P}\ \mathbb{N}$$

$$\mathbb{N}_1 = \mathbb{N} \setminus \{0\}$$

All the built-in operators and sets of Z are defined in a similar way.

The standard operators on natural numbers are defined in Z using a schema framework. There are a number of important points to notice about these schemas. First, the operators are defined as functions using the lambda notation. Second, the schemas shown illustrate how binary operators can be defined in Z. A binary operator is written with underscores on either side of the operator, both in the signature and predicate part of the schema. For example, the addition operator $+$ is defined as a total function

$$\_+\_ : \mathbb{N} \times \mathbb{N} \rightarrow \mathbb{N}$$

in the signature of the schema which follows, and is defined as a lambda expression

$$\_+\_ = \lambda\, m, n : \mathbb{N} \bullet succ^n\, m$$

in the predicate of its schema where *succ* is the successor function.

$$\begin{array}{|l}
_+_, _*_ : \mathbb{N} \times \mathbb{N} \rightarrow \mathbb{N} \\
_-_ : \mathbb{N} \times \mathbb{N} \mathrel{+\!\!\!\rightarrow} \mathbb{N} \\
\leq : \mathbb{N} \leftrightarrow \mathbb{N} \\
\hline
_+_ = \lambda\, m, n : \mathbb{N} \bullet succ^n\, m \\
_*_ = \lambda\, m, n : \mathbb{N} \bullet (_+m)^{n-1} m \\
_-_ = \lambda\, m, n : \mathbb{N} \bullet pred^n\, m \\
\leq = succ^*
\end{array}$$

$+$ thus defines the function

$$\{((0,0),0),((0,1),1),((1,1),2),\ldots\}$$

which takes an ordered pair and delivers the sum of the numbers in the pair. This enables the specifier to curry such functions to construct total functions which contain one argument. For example, the function

$$(\_+3)$$

adds three to its argument.

This is used in the definition of $*$ where multiplication is defined as repeated addition since

$$(\_+m)$$

is a function which adds m to its argument and the $(n-1)$-fold composition of the function

$$(\_+m)^{n-1}$$

when applied to m forms $m * n$ by adding m to itself $(n-1)$ times. Less than or equal $\leq$ is defined by the reflexive transitive closure of the function *succ*.

The relational operators $>$, $<$, and $\geq$ are not formally defined here. They can be built up from $\leq$. For example, $>$ is the inverse of $\leq$.

A set of consecutive natural numbers can be defined in Z using the $..$ operator. The full definition follows,

Name

$$\_ \mathinner{..} \_ : \mathbb{N} \times \mathbb{N} \rightarrow \mathbb{P}\ \mathbb{N}$$

$$m \mathinner{..} n = \{ i : \mathbb{N} \mid m \leq i \land i \leq n \}$$

9.3 Sets

All the standard operators on sets are defined in Z. Difference, union, intersection are defined as follows:

$[T]$

$$\_ \setminus \_, \_ \cup \_, \_ \cap \_ : \mathbb{P}\ T \times \mathbb{P}\ T \rightarrow \mathbb{P}\ T$$

$$\begin{aligned} &\forall S_1, S_2 : \mathbb{P}\ T \bullet \\ &\quad S_1 \setminus S_2 = \{ x : T \mid x \in S_1 \land x \notin S_2 \} \land \\ &\quad S_1 \cup S_2 = \{ x : T \mid x \in S_1 \lor x \in S_2 \} \land \\ &\quad S_1 \cap S_2 = \{ x : T \mid x \in S_1 \land x \in S_2 \} \end{aligned}$$

This definition differs from previous ones in two ways. First, double lines are used and, second, a type is enclosed in square brackets within the double lines. This is a form of schema used for **generic definitions**. A generic definition is a definition of a function, set, relation, or sequence where the type which is used is irrelevant. For example, the definition above describes three set operators which can be used for any type: sets of integers, natural numbers, chemical rectors, monitors, etc. A generic definition effectively parametrizes the schema over any type.

The definition of the subset and proper subset relations is shown below. Again it is generic.

$[T]$

$$\_ \subset \_, \_ \subseteq \_ : \mathbb{P}\ T \leftrightarrow \mathbb{P}\ T$$

$$\begin{aligned} &\forall S_1, S_2 : \mathbb{P}\ T \bullet \\ &\quad S_1 \subseteq S_2 \Leftrightarrow (\forall x : T \bullet x \in S_1 \Rightarrow x \in S_2) \land \\ &\quad S_1 \subset S_2 \Leftrightarrow (S_1 \neq S_2 \land S_1 \subseteq S_2) \end{aligned}$$

Two operators which have not been defined before are distributed union $\bigcup$ and distributed intersection $\bigcap$, also known as generalized union and intersection. The generalized union of a power set of a particular type is the union of all the elements of the sets contained in the set which is its argument. For example

$$\bigcup\{\{1,2\},\{1,2,3,4\},\{1,2,5,16,78\}\}$$

is the set

$$\{1,2,3,4,5,16,78\}$$

The generalized intersection of a power set of a particular type is the intersection of all the elements of the sets contained in the set which is its argument. For example,

$$\bigcap\{\{1,2\},\{1,2,3,4\},\{1,2,5,16,78\}\}$$

is the set

$$\{1,2\}$$

and

$$\bigcap\{\{1,4,45\},\{2,34,89\},\{203,888\}\}$$

is the empty set.

Their definitions are shown below. They are functions which have a power set of a power set as their domain and a power set as their range.

$$
\begin{array}{|l}
[T] \\
\hline
\bigcup\_, \bigcap\_ : \mathbb{P}(\mathbb{P}\ T) \rightarrow \mathbb{P}\ T \\
\hline
\forall A : \mathbb{P}(\mathbb{P}\ T) \bullet \\
\quad \bigcup A = \{ x : T \mid \exists a : A \bullet x \in a \} \land \\
\quad \bigcap A = \{ x : T \mid \forall a : A \bullet x \in a \} \\
\hline
\end{array}
$$

These are unary operators so that only one underscore is used in their definition.

Worked example 9.1 Write down a generic schema which defines a binary operator $\square$ that takes two sets and returns the cardinality of their intersection.

Solution The schema is shown below.

$$
\begin{array}{|l}
[T] \\
\hline
\_\square\_ : \mathbb{P}\ T \times \mathbb{P}\ T \rightarrow \mathbb{N} \\
\hline
\forall S_1, S_2 : \mathbb{P}\ T \bullet \\
\quad S_1 \square S_2 = \#(S_1 \cap S_2) \\
\hline
\end{array}
$$

Notice that there is no need to define formally $\cap$ within the schemas as it can be assumed that it has already been defined elsewhere in the toolbox.
■

Worked example 9.2 Define a binary operator $\mho$ which takes as its first operand a set of natural numbers and as its second operand a natural number. Its result is the number of natural numbers in the first operand which are less than the second operand.

Solution The definition is shown below.

$$\_\mho\_ : \mathbb{P}(\mathbb{N}) \times \mathbb{N} \rightarrow \mathbb{N}$$

$$\forall S_1 : \mathbb{P}\ \mathbb{N};\ s : \mathbb{N} \bullet \\ \quad S_1 \mho s = \#\{\ n : \mathbb{N} \mid n \in S_1 \wedge n < s\ \}$$

Notice that since the definition is not generic the axiomatic specification form of the schema is used.
■

Before describing relations in Z it is worth describing a facility in the language which has not been described in the first part of this book. If a set is finite, that is its members can be counted, then the symbol $\mathbb{F}$ is used. Thus, if *cars* is the set of cars then $\mathbb{F}$ *cars* describes all the finite subsets of *cars*. Since $\mathbb{P}\ S$ is the set of all subsets of S we have

$$\mathbb{F}\ S \subseteq \mathbb{P}\ S$$

9.4 Relations

Relations can also be used in Z specifications. A relation between two types T_1 and T_2 is written in the signature part of a Z schema as

$$T_1 \leftrightarrow T_2$$

Thus, a relation *HasBlocks* over *programs* and subsets of *BlockNos* would have a signature

$$\mathit{HasBlocks} : \mathit{programs} \leftrightarrow \mathbb{P}\ \mathit{BlockNos}$$

The identity relation of a set S whose elements are of type T is described in Z as

$$\mathrm{id}\, X == \{\ x : X \bullet (x, x)\ \}$$

This states that id is a function which, when it operates on a type X, delivers all the pairs of X where the first element of the pair is equal to the second element of the pair. The symbol $\mapsto$ is normally used in Z documents to stand for an ordered pair. Thus,

$$\{1 \mapsto 2, 2 \mapsto 3, 5 \mapsto 6, 7 \mapsto 1\}$$

is equivalent to

$$\{(1,2),(2,3),(5,6),(7,1)\}$$

All the operators described in Chapter 6 can be defined in Z. Relational composition can be defined as

$$
\begin{array}{l}
[T_1, T_2, T_3] \\
\_ ⨾ \_ : (T_1 \leftrightarrow T_2) \times (T_2 \leftrightarrow T_3) \rightarrow (T_1 \leftrightarrow T_3) \\
\hline
R_1 ⨾ R_2 = \{\ t_1 : T_1;\ t_3 : T_3 \mid \exists\, t_2 : T_2 \bullet \\
\qquad (t_1, t_2) \in R_1 \land (t_2, t_3) \in R_2\ \}
\end{array}
$$

The reverse or backward composition operator can also be defined in Z. Its definition is shown below.

$$
\begin{array}{l}
[T_1, T_2, T_3] \\
\_ \circ \_ : (T_1 \leftrightarrow T_2) \times (T_2 \leftrightarrow T_3) \rightarrow (T_1 \leftrightarrow T_3) \\
\hline
R_1 \circ R_2 = R_2 ⨾ R_1
\end{array}
$$

Domain and range restriction operators can be similarly defined in Z.

$$
\begin{array}{l}
[T_1, T_2] \\
\_ \lhd \_ : \mathbb{P}\ T_1 \times (T_1 \leftrightarrow T_2) \rightarrow (T_1 \leftrightarrow T_2) \\
\_ \rhd \_ : (T_1 \leftrightarrow T_2) \times \mathbb{P}\ T_2 \rightarrow (T_1 \leftrightarrow T_2) \\
\hline
\forall S : \mathbb{P}\ T_1;\ R : T_1 \leftrightarrow T_2 \bullet \\
\qquad S \lhd R = \{\ t_1 : T_1;\ t_2 : T_2 \mid t_1 \in S \\
\qquad\qquad \land (t_1 \mapsto t_2) \in R \bullet t_1 \mapsto t_2\ \} \\
\forall R : T_1 \leftrightarrow T_2;\ T : \mathbb{P}\ T_2 \bullet \\
\qquad R \rhd T = \{\ t_1 : T_1;\ t_2 : T_2 \mid (t_1 \mapsto t_2) \in R \\
\qquad\qquad \land t_2 \in T \bullet t_1 \mapsto t_2\ \}
\end{array}
$$

$$
\begin{array}{l}
[T_1, T_2] \\
\_ \lhd \_ : \mathbb{P}\ T_1 \times (T_1 \leftrightarrow T_2) \rightarrow (T_1 \leftrightarrow T_2) \\
\_ \rhd \_ : (T_1 \leftrightarrow T_2) \times \mathbb{P}\ T_2 \rightarrow (T_1 \leftrightarrow T_2) \\
\hline
\forall S : \mathbb{P}\ T_1;\ R : T_1 \times T_2 \bullet \\
\quad S \lhd R = \{\ t_1 : T_1;\ t_2 : T_2 \mid t_1 \notin S \\
\qquad \wedge (t_1 \mapsto t_2) \in R \bullet t_1 \mapsto t_2\ \} \\
\forall R : T_1 \leftrightarrow T_2;\ T : \mathbb{P}\ T_2 \bullet \\
\quad R \rhd T = \{\ t_1 : T_1;\ t_2 : T_2 \mid (t_1 \mapsto t_2) \in R \\
\qquad \wedge t_2 \notin T \bullet t_1 \mapsto t_2\ \}
\end{array}
$$

The relational image operator is similarly defined.

$$
\begin{array}{l}
[T_1, T_2] \\
\_(\!|\_|\!) : (T_1 \leftrightarrow T_2) \times \mathbb{P}\ T_1 \rightarrow \mathbb{P}\ T_2 \\
\hline
\forall R : T_1 \leftrightarrow T_2;\ S : \mathbb{P}\ T_1 \bullet \\
\quad R(\!|S|\!) = \{\ t_1 : T_1;\ t_2 : T_2 \mid t_1 \in S \wedge (t_1 \mapsto t_2) \in R \bullet t_2\ \}
\end{array}
$$

The definition of iteration and transitive closure is also shown below.

$$
\begin{array}{l}
[T_1] \\
iter : \mathbb{Z} \rightarrow (T_1 \leftrightarrow T_1) \rightarrow (T_1 \leftrightarrow T_1) \\
\hline
\forall R : T_1 \leftrightarrow T_1 \bullet \\
\quad iter\ 0\ R = \mathrm{id}\ T_1 \wedge \\
\quad (\forall k : \mathbb{N} \bullet iter(k+1)R = R \mathbin{;} (iter\ k\ R)) \wedge \\
\quad (\forall k : \mathbb{N} \bullet iter(-k)R = iter\ k\ (R^{-1})
\end{array}
$$

An iteration of a relation R to n can be written as n *iter* R or, more commonly R^n. Given this definition of *iter* the transitive closure operators can now be defined.

$$
\begin{array}{l}
[T_1] \\
\_^*, \_^+ : (T_1 \leftrightarrow T_1) \rightarrow (T_1 \leftrightarrow T_1) \\
\hline
R^* = \bigcup\{\ n : \mathbb{N} \bullet iter\ n\ R\ \} \\
R^+ = \bigcup\{\ n : \mathbb{N}_1 \bullet iter\ n\ R\ \}
\end{array}
$$

Worked example 9.3 Using the generic notation define the ran and dom operators which extract the range and domain of a relation.

Solution The schema is shown below.

$$
\begin{array}{|l}
\hline
[T_1, T_2] \\
\mathrm{dom} : (T_1 \leftrightarrow T_2) \rightarrow \mathbb{P}\ T_1 \\
\mathrm{ran} : (T_1 \leftrightarrow T_2) \rightarrow \mathbb{P}\ T_2 \\
\hline
\forall R : T_1 \leftrightarrow T_2 \bullet \\
\quad \mathrm{dom}\, R = \{\ t_1 : T_1;\ t_2 : T_2 \mid (t_1 \mapsto t_2) \in R \bullet t_1\ \} \land \\
\quad \mathrm{ran}\, R = \{\ t_1 : T_1;\ t_2 : T_2 \mid (t_1 \mapsto t_2) \in R \bullet t_2\ \} \\
\hline
\end{array}
$$

■

Worked example 9.4 A file store of a computer is modelled by means of a relation over *users* and *files*. The static properties of the file store are embodied in the schema *FileSys* shown below

$$
\begin{array}{|l}
\hline
\textit{FileSys} \\
\textit{FileStore} : \textit{USERS} \leftrightarrow \textit{FILES} \\
\textit{SystemUsers} : \mathbb{F}\ \textit{USERS} \\
\hline
\#\,\mathrm{ran}\ \textit{FileStore} \leq 1000 \\
\forall\, us : \textit{USERS} \bullet \\
\quad \#(\textit{FileStore}\,(\!|\{us\}|\!)) \leq 50 \\
\#\textit{SystemUsers} \leq 150 \\
\hline
\end{array}
$$

Using natural language describe the properties embodied in the shcema.

Solution The first predicate states that there may be no more than 1000 entries in the file store. This limits the number of files in the store. The second predicate states that for all users the number of pairs with the user's name as the first element must be less than or equal to 50. This limits the number of files which can be owned by an individual user to no more than 50. The third predicate states that there will be no more than 150 users of the system.

■

Worked example 9.5 Define a Z schema which describes the event which occurs when adding a set of user files *FilesAdd*? to the file store descibed in the previous worked example. The name of the user whose files are to be added is *us*? Assume that the schema is called *AddFiles* and has both *FileSys* and *FileSys′* included in it.

Solution The schema is shown below.

$$
\begin{array}{|l}
\hline
\textit{AddFiles} \\
\textit{FileSys} \\
\textit{FileSys}' \\
\textit{us}? : \textit{USERS} \\
\textit{FilesAdd}? : \mathbb{F}\ \textit{FILES} \\
\hline
\textit{us}? \in \textit{SystemUsers} \\
\forall f : \textit{FilesAdd}? \bullet (\textit{us}? \mapsto f) \notin \textit{FileStore} \\
\textit{FileStore}' = \textit{FileStore} \cup \{ \textit{fa} : \textit{FilesAdd}? \bullet (\textit{us}? \mapsto \textit{fa}) \} \\
\hline
\end{array}
$$

■

9.5 Functions and sequences

9.5.1 Functions

Since functions are only a special type of relation the set and relational operators described in the previous subsection can all be used with Z objects that are described as functions. The only operator which needs to be described is the functional overriding operator:

$$
\begin{array}{|l}
\hline
[T_1, T_2] \\
\_ \oplus \_ : (T_1 \twoheadrightarrow T_2) \times (T_1 \twoheadrightarrow T_2) \to (T_1 \twoheadrightarrow T_2) \\
\hline
\forall f, g : (T_1 \twoheadrightarrow T_2) \bullet \\
\quad f \oplus g = ((\mathrm{dom}\ g) \lhd f) \cup g \\
\hline
\end{array}
$$

All the symbols introduced in Chapter 7 are used to introduce functions, for example $\twoheadrightarrow$ is used to introduce a partial function.

Whenever a function is written in a signature, its type should be written using one of the functional symbols. For example, the Schema *MonSys* which follows describes the static or invariant properties of a chemical plant monitoring system. Each monitoring computer is attached to a named cluster of monitoring instruments such as thermocouples or pressure sensors. Furthermore, each monitoring instrument is attached to a particular reactor, although one reactor may have a number of instruments attached.

$$\begin{array}{|l}
\textit{MonSys} \\
\hline
attached : COMPUTERS \rightarrowtail\!\!\!\!\rightarrow clusters \\
connected : clusters \rightarrowtail\!\!\!\!\rightarrow \mathbb{F}\, INSTRUMENTS \\
monitors : INSTRUMENTS \twoheadrightarrow REACTORS \\
MaxClust, MaxComps : \mathbb{N} \\
\hline
\#\operatorname{dom} attached \leq MaxComps \\
\forall x : \operatorname{dom} connected \bullet \#connected\ x \leq MaxClust \\
\forall x, y : \operatorname{ran} connected \bullet x \neq y \Rightarrow x \cap y = \varnothing \\
\bigcup \operatorname{ran} connected = \operatorname{dom} monitors \\
\operatorname{ran} attached = \operatorname{dom} connected \\
\hline
\end{array}$$

COMPUTERS is a basic set of all possible computer names, *clusters* is the set of all possible names of clusters of instruments, *REACTORS* is a basic set of all possible computer names, and *INSTRUMENTS* is the set of all possible instrument names.

attached, *connected*, and *monitors* are all functions since computers are only connected to one cluster, clusters are attached to a unique set of instruments, and each instrument is attached to only one reactor. Moreover, *attached* is an injective function since one cluster is connected to only one computer; *connected* is also an injective function since each distinct set of instruments is connected to only one cluster. Note that *monitors* is not injective because one reactor may have any number of instruments attached to it.

There are five predicates in *MonSys* which express the static properties of the monitoring system. The first asserts that the number of computers will never be more than *MaxComps*; the second asserts that the maximum number of instruments in any one cluster will be *MaxClust*; the third asserts that no instrument will be connected in two or more clusters; the fourth asserts that all the clusters that are attachable to instruments are connected to a computer.

Finally, the fifth predicate asserts that all the clusters that are attached to instruments are connected to a computer.

Worked example 9.6 The computer system for monitoring the progress of salesmen in a computer company is to be developed. Each salesman reports to a sales manager in the company who, in turn, reports to one of two sales directors. For efficiency reasons no more than 10 salesmen report to a sales manager. Each salesman is responsible for up to 50 customers. No salesman reports to more than one sales manager, no sales manager reports to more than one sales director, and no customer is visited by more than one salesman. Construct a schema *sales* which describes the static properties of the system. Assume that the signature of the schema includes:

$$
\begin{array}{l}
DirResponsible : directors \pinj \mathbb{P}\ managers \\
ManResponsible : managers \pinj \mathbb{P}\ salesmen \\
SalesResponsible : salesmen \pinj \mathbb{P}\ customers
\end{array}
$$

Solution The schema is shown in Figure 9.1 where *directors* is the set of all

$$
\begin{array}{|l}
\textit{SalesState} \\
\hline
DirResponsible : directors \pinj \mathbb{P}\ managers \\
ManResponsible : managers \pinj \mathbb{P}\ salesmen \\
SalesResponsible : salesmen \pinj \mathbb{P}\ customers \\
DirectorStaff : \mathbb{F}\ directors \\
ManagerialStaff : \mathbb{F}\ managers \\
SalesStaff : \mathbb{F}\ salesmen \\
CurrentCustomers : \mathbb{F}\ customers \\
\hline
\forall dir : \mathrm{dom}\ DirResponsible \bullet \\
\quad DirResponsible\ dir \subseteq ManagerialStaff \\
\\
\forall man : \mathrm{dom}\ ManResponsible \bullet \\
\quad \#ManResponsible\ man \leq 10 \land \\
\quad ManResponsible\ man \subseteq SalesStaff \\
\\
\forall sm : \mathrm{dom}\ SalesResponsible \bullet \\
\quad \#SalesResponsible\ sm \leq 50 \land \\
\quad SalesResponsible\ sm \subseteq CurrentCustomers \\
\\
\#DirectorStaff = 2 \\
\\
\forall x, y\ \mathrm{ran}\ ManResponsible \bullet x \neq y \Rightarrow x \cap y = \varnothing \\
\\
\forall x, y\ \mathrm{ran}\ SalesResponsible \bullet x \neq y \Rightarrow x \cap y = \varnothing \\
\\
\forall x, y\ \mathrm{ran}\ DirResponsible \bullet x \neq y \Rightarrow x \cap y = \varnothing \\
\hline
\end{array}
$$

Figure 9.1 The schema *SalesState*

possible director names, *managers* is the set of all possible manager names, *salesmen* is the set of all possible salesman names, and *customers* is the set of customers who are visited by salesmen. All the functions are partial since not every value from *directors*, *managers*, *salesmen*, and *companies* will occur in their domains. All the functions are injective since each element in the range of the functions is associated with only one element of

the domain. Notice, also, that the sets are finite. This will normally be the case with the vast majority of sets used in the specifications of real systems.
■

9.5.2 Sequences

Since sequences play an important part in any computer system, Z has notational facilities for representing sequences and a series of operators which take sequences as operands. There are three types of sequence in Z: finite sequences, non-empty finite sequences, and injective sequences. The second type of sequence contains at least one member while the third type of sequence will contain no repetitions. They are defined as

$$\begin{array}{l} \mathrm{seq}\ T == \{\ f : \mathbb{N} \twoheadrightarrow\!\!\!\!\!\!\!+\ \ T \mid \mathrm{dom} f = 1 \mathinner{..} \# f\ \} \\ \mathrm{seq}_1\ T == \{\ f : \mathrm{seq}\ T \mid \# f > 0\ \} \\ \mathrm{iseq}\ T == \mathrm{seq}\ T \cap (\mathbb{N} \rightarrowtail\!\!\!\!\!\!+\ \ T) \end{array}$$

A sequence then is a partial function which maps natural numbers from 1 up to the cardinality of the function, into objects which have a type T. Thus,

$$\{1 \mapsto 3, 2 \mapsto 9, 3 \mapsto 9, 4 \mapsto 11\}$$

is a sequence of natural numbers ($\mathrm{seq}\,\mathbb{N}$) and

$$\{1 \mapsto \mathit{UpdateFile}, 2 \mapsto \mathit{edits}, 3 \mapsto \mathit{NewTaxFile}\}$$

is an example of an injective sequence of files ($\mathrm{iseq}\,\mathit{files}$). The length of a sequence s in Z is given by $\#s$. In Z the normal way to express sequences is to use angle brackets to enclose the elements of a sequence; for example, the sequence

$$\langle \mathit{UpdateFile}, \mathit{edits}, \mathit{NewTaxFile} \rangle$$

is equivalent to the preceding sequence which was expressed using the $\mapsto$ notation.

When sequences are included in the signature part of a schema, the type of the elements which make up the sequence are preceded by the letters seq. For example, the schema *FileQueue*

```
┌─ FileQueue ──────────────────────────────
│ InQueue, OutQueue : seq Files
├──────────
│ #InQueue < #OutQueue
└──────────────────────────────────────────
```

declares two objects which are sequences of files and states that the length of the first sequence *InQueue* is less than that of *OutQueue*.

The concatenation of sequences is achieved by means of the $\frown$ operator. It takes two operands which are sequences and joins the second operand to the end of the first operand. Thus, if a and b are sequences of any type then $a \frown b$ adds b to the end of a. The $\frown$ operator can also be used to add single elements to the beginning or end of a queue. Thus, if a is a queue whose elements are of type T and t is of type T, then

$$\langle t \rangle \frown a$$

adds t to the front of a and

$$a \frown \langle t \rangle$$

adds t to the end of a. The formal definition of concatenation is shown below.

$$[T]$$
$$\_ \frown \_ : \text{seq}\ T \times \text{seq}\ T \rightarrow \text{seq}\ T$$
$$\forall\, s, t : \text{seq}\ T \bullet$$
$$s \frown t = s \cup \{\ n : \text{dom}\, t \bullet n + \#s \mapsto t\ n\ \}$$

As an example of the use of queues, consider the device handler for a moving-head disc unit. This is activated by a series of read/write instructions. These instructions give the address of the data to be read or written together with the address of a fixed area of store from which data is to be written to or read from. Since disc access will always take a longer time than store access, there will almost invariably be a number of requests waiting to be satisfied.

The device handler can be modelled by means of two sequences. The first would hold read requests; the second would hold write requests. The elements of the queue will be pairs consisting of a disc address and a memory address. Thus, the queues can be represented in a signature as

$$\mathit{ReadQueue}, \mathit{WriteQueue} : \text{seq}(\mathit{DiscAddresses} \times \mathit{MemoryAddresses})$$

where the variable *DiscAddresses* is the set of all possible disc addresses and the variable *MemoryAddresses* is the set of all possible memory addresses.

Sequence selection operators can be defined in Z as follows:

$$
\begin{array}{|l}
\multicolumn{1}{l}{=[T]=} \\
head, last : \mathrm{seq}_1\ T \rightarrow T \\
tail, front : \mathrm{seq}_1\ T \rightarrow \mathrm{seq}\ T \\
\hline
\forall s : \mathrm{seq}_1\ T \bullet \\
\quad head\ s = s(1) \land \\
\quad last\ s = s(\#s) \land \\
\quad tail\ s = (\lambda\, n : 1 \mathrel{..} \#s - 1 \bullet s(n+1)) \land \\
\quad front\ s = (1 \mathrel{..} \#s - 1) \lhd s \\
\hline
\end{array}
$$

head returns the first element from a sequence, *tail* returns all but the first element of a sequence, *front* returns all but the last element of a sequence, and *last* returns the last element of a sequence. The fact that the domain of these functions is a set of sequences which contain one element ensures that the functions are defined for non-empty lists.

Worked example 9.7 During the operation of a computer operating system, users will be continually creating and deleting files. In a distributed operating system the files that are to be deleted or created may be held on a computer that is remote from the one on which the user is working. The file store of each computer will contain blocks of storage which are currently used in files and blocks of storage which are free for newly created files.

During the operation of the distributed system a packet of blocks released from the deletion of files by users of other computers will be received by a computer together with blocks released by users of that computer. The blocks will be queued in the order in which they have been received. There will be no more than 30 computers in the distributed system. How would you model the distributed file store together with the queues of released blocks? Ignore any requirements to model the creation of files. Write down a schema which describes the static properties of the system; assume that the two sets *BLOCKS* and *COMPUTERS* are basic sets.

Solution The file store of each computer can be modelled by means of a disjoint set of blocks. The first set contains those blocks in use; the second set contains those blocks available for new files.

Each computer will be associated with a queue of blocks. These queues can be modelled by means of a sequence of sets of blocks. The schema which describes the static properties of the system is shown in Figure 9.2 The free blocks in each computer together with the used blocks for each computer are modelled by the partial functions *FreeFileStores* and *UsedFileStores*. The queues of blocks waiting to be relinquished to the free block store of each computer are modelled by means of the partial function *WaitingDelete*.

■

$[COMPUTERS, BLOCKS]$

$$
\begin{array}{l}
\underline{\quad CompNet \quad} \\
ConnectedComputers : \mathbb{F}\ COMPUTERS \\
FreeFileStores : COMPUTERS \twoheadrightarrow\!\!\!\!\!\!\!\;+ \ \mathbb{F}\ BLOCKS \\
UsedFileStores : COMPUTERS \twoheadrightarrow\!\!\!\!\!\!\!\;+ \ \mathbb{F}\ BLOCKS \\
WaitingDelete : COMPUTERS \twoheadrightarrow\!\!\!\!\!\!\!\;+ \ \mathrm{seq}(\mathbb{F}\ BLOCKS) \\
\hline
ConnectedComputers = \mathrm{dom}\ FreeFileStores \\
\mathrm{dom}\ FreeFileStores = \mathrm{dom}\ UsedFileStores \\
\mathrm{dom}\ FreeFileStores = \mathrm{dom}\ WaitingDelete \\
\#ConnectedComputers \leq 30 \\
\forall\ comp : ConnectedComputers \bullet \\
\quad FreeFileStores\ comp \cap UsedFileStores\ comp = \varnothing \\
\forall\ comp : ConnectedComputers \bullet \\
\quad \forall\ x, y : \mathrm{ran}\ WaitingDelete\ comp \bullet \\
\qquad x \neq y \Rightarrow x \cap y = \varnothing \\
\forall\ comp : ConnectedComputers \bullet \\
\quad \forall\ x : \mathrm{ran}\ WaitingDelete\ comp \bullet \\
\qquad x \subseteq UsedFileStores\ comp
\end{array}
$$

Figure 9.2 The schema *CompNet*

Worked example 9.8 Write down the schema *ReturnBlock* which describes the event of returning the first set of blocks of file storage to free storage from the queue of returned blocks associated with the computer $cn?$

Solution The schema is shown below. The first predicate states that the computer $cn?$ must be one of the connected computers. The second predicate states that the queue must not be empty. The third predicate states that the operation does not alter the number or names of the computers in the distributed system. The fourth predicate describes the addition of the queued block numbers to the free store associated with the computer $cn?$ The fifth predicate describes the removal of the queued block numbers from the used store associated with the computer $cn?$ Finally, the sixth predicate describes the removal of the first element from the queue associated with the computer $cn?$

$$
\begin{array}{|l}
\hline
\textit{ReturnBlock} \\
\textit{CompNet} \\
\textit{CompNet}' \\
cn? : \textit{COMPUTERS} \\
\hline
cn? \in \textit{ConnectedComputers} \\
\textit{WaitingDelete}\ cn? \neq \langle\,\rangle \\
\textit{ConnectedComputers}' = \textit{ConnectedComputers} \\
\textit{FreeFileStores}' = \\
\quad \textit{FreeFileStores} \oplus \\
\quad \{cn? \mapsto \textit{FreeFileStores}\ cn? \cup \textit{head}(\textit{WaitingDelete}\ cn?)\} \\
\textit{UsedFileStores}' = \\
\quad \textit{UsedFileStores} \oplus \\
\quad \{cn? \mapsto \textit{UsedFileStores}\ cn? \setminus \textit{head}(\textit{WaitingDelete}\ cn?)\} \\
\textit{WaitingDelete}' = \textit{WaitingDelete} \oplus \\
\quad \{cn? \mapsto \textit{tail}(\textit{WaitingDelete}\ cn?)\} \\
\hline
\end{array}
$$

■

To conclude this section some other sequence facilities are described. The functions *for* and *after* retrieve the first n elements of a sequence and the last n elements.

$$
\begin{array}{|l}
\hline
[T] \\
\textit{after}, \textit{for} : (\mathrm{seq}\ T) \times \mathbb{N} \rightarrow \mathrm{seq}\ T \\
\hline
\forall s : \mathrm{seq}\ T;\ n : \mathbb{N} \bullet \\
\quad s\ \textit{after}\ n = \textit{succ}^{\#s-n} \mathbin{;} s \wedge \\
\quad s\ \textit{for}\ n = (1 \mathbin{..} n) \lhd s \\
\hline
\end{array}
$$

The final facility is the function rev which takes a sequence and reverses its elements. Thus if the sequence sq is

$$\langle 6, 22, 1, 14 \rangle$$

then rev sq is

$$\langle 14, 1, 22, 6 \rangle$$

The definition is

$$
\begin{array}{l}
[T] \\
\hline
\text{rev} : \text{seq}\ T \rightarrow \text{seq}\ T \\
\hline
\forall s : \text{seq}\ T \bullet \\
\quad \text{rev}\ s = (\lambda\, n : \text{dom}\, s \bullet s(\#s - n + 1)) \\
\hline
\end{array}
$$

9.6 Modelling a back order system

This section concludes the description of Z basic facilities. It illustrates the use of many of these facilities in the specification of part of a typical commercial data processing system. The specification is to be produced from the statement of requirements.

9.6.1 The statement of requirements

The statement of requirements is shown below.

1. The back order subsystem is to provide facilities for the processing of orders for commodities associated with gardening which are received by a wholesaler and which cannot be immediately satisfied by the wholesaler from stock.

2. On each day of operation a number of orders are received by the wholesaler. The majority of these orders are normally satisfied by withdrawing items from stock. However, some items will be temporarily out of stock.

3. When an order for an article is received which cannot be satisfied, the order is placed on a queue of orders for that article. The place where the order is inserted in the queue will depend on the importance of the customer who places the order. Some customers, for example, large chain stores, are regarded as more important than other customers, so a priority list of customers must be maintained.

4. A series of commands should be provided as part of the back order subsystem. They are described in the following paragraphs.

5. NUMBERQUEUE. This command displays the reference number of each customer who has ordered a quantity of a commodity that is out of stock. The customers, together with their associated order number, are displayed in the same order in which they occur in the commodity queue. The enquiries clerk who types in the command will provide the reference number of the commodity.

6. TOTALQUEUE. This command displays the total number of orders for a particular out-of-stock commodity. The enquiries clerk who types in the command will provide the reference number of the commodity.

7. TOTALALLQUEUE. This command displays the total number of commodities which are currently ordered but are out of stock.

8. INSERTCUSTOMER. The subsystem shall keep account of each possible customer for commodities. Associated with each customer is a priority. Customers with a high priority will have their orders satisfied before customers with a low priority. The purpose of this command is to insert a new customer into the system together with an associated priority. The order clerk types in a customer number and a priority.

9. DELETECUSTOMER. This command removes a single customer from the list of customers. The order clerk types in the customer number.

10 MOVECUSTOMER. This command alters the priority of a customer. The order clerk types in the customer number and a new priority.

11. REMOVEPRODUCT. Very occasionally the wholesaler is notified by a manufacturer that a particular commodity is no longer produced. This command removes such commodities from the subsystem. The order clerk types in the commodity number that is no longer produced.

12. NEWPRODUCT. Whenever a manufacturer decides to produce a new commodity, the number of that commodity is added to the subsystem. The order clerk types in the number of the new commodity.

13. NUMBERSTOCKED. This command displays the number of commodities that are currently stocked.

Should the priorities of the customers be unique? If so, the position of individual orders belonging to a customer is unambiguously fixed in relation to orders from other customers. However, the ordering of orders from the same customer within the back order queues is not specified. We shall assume that when a series of orders from the same customer are in a back order queue, they are placed in ascending order of quantity. This has the dubious advantage that, when a consignment of a product is received by the wholesaler, then a number of smaller orders may be satisfied rather than one or two larger orders.

What should happen if a customer is removed from those customers currently known to the subsystem by the DELETECUSTOMER command and a number of back orders for that customers are stored on a back order queue? Obviously, the statement of requirements is incomplete. Something will have to be done as a back order will normally correspond to an order to manufacturer which may have to be cancelled. In this case a print-out of orders from the deleted customer should be produced.

What would happen when the REMOVEPRODUCT command is typed? There may be some orders for this product which can never be satisfied. It is clear that the customers for this product who have orders on the back order queue will need to be notified or a substitute product sent. The statement of requirements is thus incomplete. We shall assume that a print-out of customer and order numbers for that product will be produced when that product is deleted from the subsystem.

What should happen when the MOVECUSTOMER command is typed? The effect of the command is to change the priority of a customer. Does this mean that the queue of back orders from this customer should take account of the change? For example, if the customer is given a new higher priority should his or her back orders be moved up a back order queue? The statement of requirements is incomplete in this respect. We shall assume that the back order queue will be adjusted.

9.6.2 Specifying the static properties of the system

The major part of the statement of requirements is concerned with queues of back orders. These queues can be modelled by sequences. The whole collection of queues, each queue associated with a product, can be modelled by means of a partial function that maps product numbers to sequences which model the back order queues for that product.

The statement of requirements also refers to items in stock. This means that the association between a product and a number in stock should be modelled. This is achieved by means of a partial function that maps product numbers to the number in stock of that particular product.

This means that the association between a product and a number in stock should be modelled. This is achieved by means of a partial function that maps product numbers to the number in stock of that particular product.

Finally, the subsystem will need to keep track of customer priorities. Since each customer has a unique priority, the association between a customer and a priority can be modelled by means of a partial injective function which maps customer numbers to priorities. Assuming that the basic sets are

$$[PRODUCTNOS, CUSTNOS, PRIORITYNOS, ORDERNOS]$$

the system can be characterized by the schema signature

BackOrders

$$\begin{array}{l} queues : PRODUCTNOS \pinj \mathrm{seq}\ ORDERNOS \\ HasPriority : CUSTNOS \pinj PRIORITYNOS \\ stocks : PRODUCTNOS \pfun \mathbb{N} \\ BackOrdersNo : ORDERNOS \pfun \mathbb{N} \\ BackOrdersCust : ORDERNOS \pfun CUSTNOS \end{array}$$

queues maps *PRODUCTNOS* into the sequence of customer order numbers each of which correspond to an order for a commodity. *HasPriority* associates a customer number with the priority of the customer. *stocks* maps a product number into the quantity in stock of that product. *BackOrdersNo* maps an order number into a quantity ordered and *BackOrdersCust* maps an order number into the number of the customer who is associated with the order.

Worked example 9.9 Write down in natural language the properties of the back order subsystem which must be true throughout its operation.

Solution The list of properties is

- Each commodity that is stocked must be associated with a possibly empty back order queue.

- Each order number in a queue element corresponds to an existing back order.

- The number in stock of a commodity must always be less than the number ordered in the first element of the queue for that commodity.

- Each order number in a queue element is unique.

- The orders should be in order of customer priority.

■

The schema which describes these properties is shown in Figure 9.3. Given this schema, a number of events can be specified. Each event corresponds to a command typed in by a clerk.

The first command NUMBERQUEUE can be easily specified as the event *NumberQueue* which occurs when the command is typed.

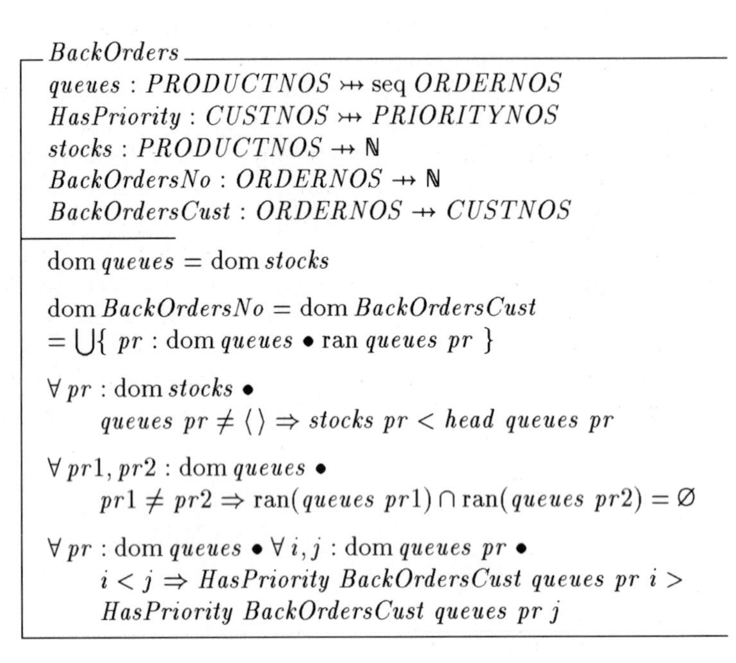

BackOrders

$queues : PRODUCTNOS \rightarrowtail\!\!\!\!\rightarrow \text{seq } ORDERNOS$
$HasPriority : CUSTNOS \rightarrowtail\!\!\!\!\rightarrow PRIORITYNOS$
$stocks : PRODUCTNOS \twoheadrightarrow\!\!\!\!\!\!\mapsto \mathbb{N}$
$BackOrdersNo : ORDERNOS \twoheadrightarrow\!\!\!\!\!\!\mapsto \mathbb{N}$
$BackOrdersCust : ORDERNOS \twoheadrightarrow\!\!\!\!\!\!\mapsto CUSTNOS$

$\text{dom } queues = \text{dom } stocks$

$\text{dom } BackOrdersNo = \text{dom } BackOrdersCust$
$= \bigcup\{\ pr : \text{dom } queues \bullet \text{ran } queues\ pr\ \}$

$\forall\, pr : \text{dom } stocks \bullet$
$\quad queues\ pr \neq \langle\rangle \Rightarrow stocks\ pr < head\ queues\ pr$

$\forall\, pr1, pr2 : \text{dom } queues \bullet$
$\quad pr1 \neq pr2 \Rightarrow \text{ran}(queues\ pr1) \cap \text{ran}(queues\ pr2) = \varnothing$

$\forall\, pr : \text{dom } queues \bullet \forall\, i, j : \text{dom } queues\ pr \bullet$
$\quad i < j \Rightarrow HasPriority\ BackOrdersCust\ queues\ pr\ i >$
$\quad HasPriority\ BackOrdersCust\ queues\ pr\ j$

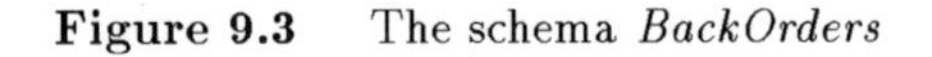

Figure 9.3 The schema *BackOrders*

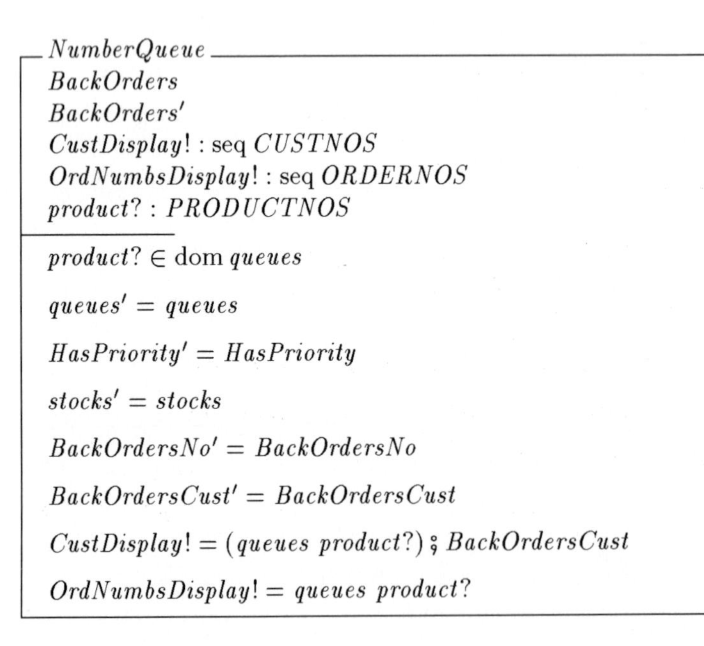

NumberQueue

$BackOrders$
$BackOrders'$
$CustDisplay! : \text{seq } CUSTNOS$
$OrdNumbsDisplay! : \text{seq } ORDERNOS$
$product? : PRODUCTNOS$

$product? \in \text{dom } queues$

$queues' = queues$

$HasPriority' = HasPriority$

$stocks' = stocks$

$BackOrdersNo' = BackOrdersNo$

$BackOrdersCust' = BackOrdersCust$

$CustDisplay! = (queues\ product?) \mathbin{;} BackOrdersCust$

$OrdNumbsDisplay! = queues\ product?$

product? is the product number typed by the clerk. Two sequences are formed: *CustDisplay!* contains the customer numbers of those customers who have back orders for *product?*, *OrdNumbsDisplay!* contains the order numbers of the back orders for *product?*. The first predicate asserts that the product is known to the subsystem. The second to sixth predicates state that the subsystem will remain unchanged by the event. The seventh and eighth predicates define the sequences *CustDisplay!* and *OrdNumbsDisplay!*

The TOTALQUEUE command is specified by a schema *TotalQueue*: it describes the event that occurs when the command is typed. *product?* represents the product number typed and *number!* represents the total number of queue entries for *product?*

The TOTALALLQUEUE command can be specified by the schema *TotalAllQueue* shown below. The first five conjuncts are straightforward; they describe the fact that the subsystem is unaffected by the event. The final conjunct is more complicated. It states that the total number of individual commodities which are on order is the cardinality of the set of products associated with back order queues which are not empty.

TotalQueue

$BackOrders\ BackOrders'$
$product? : PRODUCTNOS$
$number! : \mathbb{N}$

$product? \in \mathrm{dom}\ queues \wedge queues' = queues$

$HasPriority' = HasPriority \wedge stocks' = stocks$

$BackOrdersNo' = BackOrdersNo$

$BackOrdersCust' = BackOrdersCust$

$number! = \#(queues\ product?)$

TotalAllQueue

$BackOrders\ BackOrders'$
$TotalNumber! : \mathbb{N}$

$queues' = queues \wedge HasPriority' = HasPriority$

$stocks' = stocks \wedge BackOrdersNo' = BackOrdersNo$

$BackOrdersCust' = BackOrdersCust$

$TotalNumber! = \#\{\ prod : \mathrm{dom}\ queues \mid queues\ prod \neq \langle\,\rangle\ \}$

Worked example 9.10 Define a schema *InsertCustomer* which describes the event that occurs when the INSERTCUSTOMER command is typed.

Solution In the schema *InsertCustomer*, *customer?* is the name of the customer and *priority?* is the associated priority of the customer. The first predicate asserts that the customer must be a new customer. The second predicate asserts that the priority must be unique. The third to sixth predicates state that *queues*, *stocks*, *BackOrdersNo*, and *BackOrdersCust* are unchanged by the event. Finally, the last predicate uses the set union operator to show the addition of the pair corresponding to the new customer with a new priority to *HasPriority*.

$$
\begin{array}{|l}
\textit{InsertCustomer} \\
\hline
\textit{BackOrders} \\
\textit{BackOrders}' \\
\textit{customer?} : \textit{CUSTNOS} \\
\textit{priority?} : \textit{PRIORITYNOS} \\
\hline
\textit{customer?} \notin \operatorname{dom} \textit{HasPriority} \\
\textit{priority?} \notin \operatorname{ran} \textit{HasPriority} \\
\textit{queues}' = \textit{queues} \\
\textit{stocks}' = \textit{stocks} \\
\textit{BackOrdersNo}' = \textit{BackOrdersNo} \\
\textit{BackOrdersCust}' = \textit{BackOrdersCust} \\
\textit{HasPriority}' = \textit{HasPriority} \cup \{\ \textit{customer?} \mapsto \textit{priority?}\ \} \\
\hline
\end{array}
$$

■

The next command the DELETECUSTOMER command can be specified as the schema *DeleteCustomer* shown on the next page. It describes the event that occurs when this command is typed. The number of the customer to be deleted is *customer?* The first six predicates are relatively simple. The seventh predicate states that each of the queues after the *DeleteCustomer* event will contain the elements in the queue before the event minus those associated with the customer.

The MOVECUSTOMER command can be specified by means of the schema *MoveCustomer* shown on the next page. This describes the event of altering a customer priority. *customer?* is the name of the customer whose priority is to be changed and *NewPriority?* is the new value of the priority. The first predicate states that the customer must already have been allocated a priority number. The second predicate asserts that the new priority must not already be assigned to a current customer. The sixth predicate asserts that the new priority has been assigned. The seventh and eighth predicates assert that no new product queues are created and no existing queues are deleted.

$$
\begin{array}{l}
\textit{DeleteCustomer} \\
\hline
\textit{BackOrders}\ \textit{BackOrders}' \\
\textit{customer}? : \textit{CUSTNOS} \\
\hline
\textit{customer}? \in \operatorname{dom} \textit{HasPriority} \\
\textit{HasPriority}' = \{\textit{customer}?\} \mathbin{⩤} \textit{HasPriority} \\
\textit{stocks}' = \textit{stocks} \\
\textit{BackOrdersCust}' = \textit{BackOrdersCust} \mathbin{⩥} \{\textit{customer}?\} \\
\textit{BackOrdersNo}' = (\operatorname{dom} \textit{BackOrdersCust}') \lhd \textit{BackOrdersNo} \\
\operatorname{dom} \textit{queues}' = \operatorname{dom} \textit{queues} \\
\forall p : \operatorname{dom} \textit{queues}' \bullet \\
\quad \textit{queues}'\, p = \textit{squash}(\textit{queues}\ p \mathbin{⩥} \\
\quad \{\ o : \textit{ORDERNOS} \mid \textit{BackOrdersCust}\ o = \textit{customer}?\ \}) \\
\hline
\end{array}
$$

$$
\begin{array}{l}
\textit{MoveCustomer} \\
\hline
\textit{BackOrders}\ \textit{BackOrders}' \\
\textit{customer}? : \textit{CUSTNOS} \\
\textit{NewPriority}? : \textit{PRIORITYNOS} \\
\hline
\textit{customer}? \in \operatorname{dom} \textit{HasPriority} \\
\textit{NewPriority}? \notin \operatorname{ran} \textit{HasPriority} \\
\textit{stocks}' = \textit{stocks} \\
\textit{BackOrdersNo}' = \textit{BackOrdersNo} \\
\textit{BackOrdersCust}' = \textit{BackOrdersCust} \\
\textit{HasPriority}' = \textit{HasPriority} \oplus \{(\textit{customer}?, \textit{NewPriority}?)\} \\
\operatorname{dom} \textit{queues}' = \operatorname{dom} \textit{queues} \\
\forall \textit{prod} : \operatorname{dom} \textit{queues} \bullet \\
\quad \operatorname{ran} \textit{queues}'\ \textit{prod} = \operatorname{ran} \textit{queues}\ \textit{prod} \\
\hline
\end{array}
$$

Worked example 9.11 Write down a schema *RemoveProduct* which describes the event that occurs when the REMOVEPRODUCT command is typed and when there is stock held for the product that is to be removed

Solution The schema is shown below.

$$
\begin{array}{l}
\textit{RemoveProduct} \\
\hline
BackOrders \\
BackOrders' \\
prod? : PRODUCTNOS \\
LuckyCustomer! : CUSTNOS \\
NotifiedCustomers : \mathbb{F}\ CUSTNOS \\
\hline
stocks\ prod? > 0 \\
prod? \in \mathrm{dom}\ queues \\
HasPriority' = HasPriority \\
LuckyCustomer! = BackOrdersCust\ queues\ prod?\ 1 \\
NotifiedCustomers! = BackOrdersCust (\!| \mathrm{ran}\ queues\ prod |\!) \\
\quad \setminus LuckyCustomer! \\
queues' = \{prod?\} \lhd queues \\
stocks' = \{prod?\} \lhd stocks \\
BackOrdersNo' = (\mathrm{ran}\ queuesprod?) \lhd BackOrdersNo \\
BackOrdersCust' = (\mathrm{ran}\ queues\ prod?) \lhd BackOrdersCust
\end{array}
$$

The name of the commodity which has been discontinued is *prod*?. If there is stock of the commodity in the wholesaler's warehouse, then the first customer on the back order queue for the commodity *LuckyCustomer*! will receive the stock of that commodity.

■

Worked example 9.12 Write down a schema *NewProduct* which describes the event that occurs when the NEWPRODUCT command is typed.

Solution The schema is shown below.

$$
\begin{array}{|l}
\hline
\textit{NewProduct} \\
\hline
\textit{BackOrders} \\
\textit{BackOrders}' \\
\textit{NewProduct}? : \textit{PRODUCTNOS} \\
\hline
\textit{NewProduct}? \notin \operatorname{dom} \textit{queues} \\
\textit{HasPriority}' = \textit{HasPriority} \\
\textit{BackOrdersNo}' = \textit{BackOrdersNo} \\
\textit{BackOrdersCust}' = \textit{BackOrdersCust} \\
\textit{queues}' = \textit{queues} \cup \{\textit{NewProduct}? \mapsto \langle\,\rangle\} \\
\textit{stocks}' = \textit{stocks} \cup \{\textit{NewProduct}? \mapsto 0\} \\
\hline
\end{array}
$$

■

9.7 Further reading

When the first edition of this book was published there were no books on Z published. However, there are now three books which I would recommend anyone interested in Z to read. The first is [Spivey, 1989] which is more of a short reference manual than a Z textbook. [Potter *et al.*, 1991] is, I think, the best book written on Z so far. While it skimps the discrete mathematics it does give a very thorough and emminently readable introduction to Z. [Diller, 1990] is another good book on Z which also contains some interesting material on the prototyping of specifications. It is a little more mathematical than [Potter *et al.*, 1991].

10

The Z schema calculus

Aims

- To describe how Z schemas can be used as objects in formal specifications.
- To introduce the schema calculus.
- To show the Z schema calculus being used in some small examples.

10.1 Schemas as objects

Schemas are a convenient way of organizing the description of specifications. They have a form that replicates a pattern found throughout the discrete mathematics described in the early parts of this book. For example, the comprehensive set specification

$$\{ \, file : SysFiles \mid file \in accessed \; SuperUser \, \}$$

is reflected in the schema

```
┌─ FileExample ──────────────────────────
│ file : SysFiles
├──────────────
│ file ∈ accessed SuperUser
└────────────────────────────────────────
```

while the function

$$\lambda \, s : \mathrm{seq} \, FileUpdates \mid \#s > 3 \bullet s(1)$$

is reflected in the schema

```
┌─ Updates ──────────────────────────────
│ file : seq FileUpdates
├──────────────
│ #s > 3
└────────────────────────────────────────
```

and the predicate

$$\forall file : SysFiles;\ us : NormalUsers \mid \neg\ (file\ IsOwned\ us)$$

is reflected in the schema

```
┌─ SystemFiles ──────────────────────────────
│ file : SysFiles
│ us : NormalUsers
├──────────────
│ ¬ (file IsOwned us)
└────────────────────────────────────────────
```

This equivalence between schemas and comprehensive set specifications, predicates, and lambda abstractions means that these three entities are interchangeable with schemas. For example, the schema

```
┌─ SpecialInvoices ──────────────────────────
│ name : INames
│ SCat, NSupp : 𝔽 INames
├──────────────
│ name ∈ NSupp ∨ name ∈ SCat
└────────────────────────────────────────────
```

when enclosed in curly brackets

$$\{SpecialInvoices\}$$

is equivalent to the comprehensive set specification

$$\{\ name : INames;\ SCat, NSupp : \mathbb{F}\ INames \mid$$
$$name \in NSupp \lor name \in SCat\ \}$$

which is the set of all invoice names which are either in the set $NSupp$ or the set $SCat$.

Predicates and schemas are also interchangeable. For example, a schema such as

```
┌─ NormalInvoice ────────────────────────────
│ name : InvoiceNames
├──────────────
│ name ∈ CurrentInvoices
└────────────────────────────────────────────
```

can be preceded by a quantifier and followed by a period and a predicate. For example,

$$\forall NormalInvoice \bullet name \subseteq ResInvoices$$
$$\exists NormalInvoice \bullet name \subseteq SpecCat$$

The period is read as an implication so that the above predicates are equivalent to

$$\exists\, name : InvoiceNames \mid name \in CurrentInvoices \Rightarrow$$
$$name \subseteq SpecCat$$

$$\forall\, name : InvoiceNames \mid name \in CurrentInvoices \Rightarrow$$
$$name \subseteq ResInvoices$$

A schema can also be quantified by just writing the quantifier in front of the schema name; thus

$$\exists\, NormalInvoice$$

is equivalent to

$$\exists\, name : InvoiceNames \bullet name \in CurrentInvoices$$

and

$$\forall\, NormalInvoice$$

is equivalent to

$$\forall\, name : InvoiceNames \bullet name \in CurrentInvoices$$

Finally, a schema can form part of a lambda expression. For example, a schema such as

```
—DistributedDirectory———————————————————
| AllUsers, NormalUsers, SystemUsers : F UserNames
|——————————
| NormalUsers ∩ SystemUsers = ∅
|
| (NormalUsers ∪ SystemUsers) ⊆ AllUsers
————————————————————————————————————————
```

can be preceded by λ and followed by a period and an expression involving the variables of the signature. Thus,

$$\lambda\, DistributedDirectory \bullet \#NormalUsers$$

is a function that gives the number of normal users in *DistributedDirectory*.

The function

$$\lambda\, DistributedDirectory \bullet SystemUsers$$

returns the set of system users in a distributed directory. Such a function which gives one of the variables that make up a signature is known as a **projection function**.

Exercise 10.1

The schema *FileQueue* describes the properties of the subsystem of an operating system concerned with the off-line printing of results and reports that have been generated by terminated user programs.

$$
\begin{array}{l}
\textit{FileQueue} \\
\hline
\textit{WaitingJobs} : \textit{Operiphs} \twoheadrightarrow \operatorname{seq} \textit{PrintFiles} \\
\textit{CurrPriority} : \textit{Operiphs} \rightarrowtail\!\!\!\twoheadrightarrow \textit{PrintPriorities} \\
\textit{MaxQsize} : \mathbb{N} \\
\hline
\operatorname{dom} \textit{WaitingJobs} = \operatorname{dom} \textit{CurrPriority} \\
\forall s : \operatorname{dom} \textit{WaitingJobs} \bullet \\
\quad \# \textit{WaitingJobs}\ s \leq \textit{MaxQSize}
\end{array}
$$

Write down

(i) The function which returns the waiting jobs in the file queue.
(ii) The assertion that in the *FileQueue* subsystem there isn't a queue longer than *MaxQSize*.
(iii) The function that gives the number of output peripherals in the *FileQueue* subsystem.
(iv) The set of all *FileQueues* ...

∎

Schemas can be written using the box notation above or can be enclosed in brackets. For example, the schema

$$
\begin{array}{l}
\textit{SUsers} \\
\hline
\textit{SysUsers}, \textit{NormalUsers} : \mathbb{N} \\
\hline
\textit{SysUsers} + \textit{NormalUsers} \leq 120
\end{array}
$$

can be written as

$$
\begin{array}{l}
\textit{SUsers} \mathrel{\widehat=} \\
\quad [\textit{SysUsers}, \textit{NormalUsers} : \mathbb{N} \mid \textit{SysUsers} + \textit{NormalUsers} \leq 120]
\end{array}
$$

10.2 Extending and manipulating schemas

The preceding chapter described how schemas can be incorporated into other schemas by means of schema inclusion. This allows a much more modular form of presentation and enables a specification to be presented at a number of levels of abstraction. However, the precise semantics of schema inclusion have not yet been described. The result of including one schema within another is to form another schema which contains the union of the signatures of both schemas with the predicates of both schemas conjoined together. For example, if the schema *files* is

$$\begin{array}{|l}
\multicolumn{1}{l}{\textit{files}} \\
\hline
SysFiles, UserFiles, StoredFiles : \mathbb{F}\ FILENAMES \\
\hline
SysFiles \cup UserFiles = StoredFiles \\
SysFiles \cap UserFiles = \varnothing \\
\hline
\end{array}$$

and the schema *tapes* is

$$\begin{array}{|l}
\multicolumn{1}{l}{\textit{tapes}} \\
\hline
SysTapes, UserTapes, StoredTapes : \mathbb{F}\ TAPENAMES \\
\hline
SysTapes \cup UserTapes = StoredTapes \\
SysTapes \cap UserTapes = \varnothing \\
\hline
\end{array}$$

then

$$\begin{array}{|l}
\multicolumn{1}{l}{WholeFile} \\
\hline
files \\
tapes \\
MaxFiles, MaxTapes : \mathbb{N} \\
\hline
\#StoredFiles \leq MaxFiles \land \#StoredTapes \leq MaxTapes \\
\hline
\end{array}$$

is equivalent to the schema

$$\begin{array}{|l}
\multicolumn{1}{l}{WholeFile} \\
\hline
SysFiles, UserFiles, StoredFiles : \mathbb{F}\ FILENAMES \\
SysTapes, UserTapes, StoredTapes : \mathbb{F}\ TAPENAMES \\
MaxFiles, Maxtapes : \mathbb{N} \\
\hline
SysFiles \cup UserFiles = StoredFiles \\
SysFiles \cap UserFiles = \varnothing \\
SysTapes \cup UserTapes = StoredTapes \\
SysTapes \cap UserTapes = \varnothing \\
\#StoredFiles \leq MaxFiles \\
\#StoredTapes \leq MaxTapes \\
\hline
\end{array}$$

Any duplicated names in the schema which are joined in this way are replaced by a single instance of the name provided they are of the same type. These names must be of the same type.

A schema can be primed. This has the effect of placing a prime above each variable in both the signature and the predicate parts of a schema. Thus, if schema *cust* is

$$\begin{array}{|l}
\textit{cust} \\
\hline
\mathit{OrdinaryCust}, \mathit{PriorityCust}, \mathit{RegisteredCust} : \mathbb{F}\ \mathit{CUSTOMERS} \\
\hline
\mathit{OrdinaryCust} \cap \mathit{PriorityCust} = \varnothing \\
\mathit{OrdinaryCust} \cup \mathit{PriorityCust} = \mathit{RegisteredCust} \\
\hline
\end{array}$$

which represents the relationships between categories of customer in a commercial data processing application, then $cust'$ is

$$\begin{array}{|l}
\textit{cust}' \\
\hline
\mathit{OrdinaryCust}', \mathit{PriorityCust},' \mathit{RegisteredCust}' : \mathbb{F}\ \mathit{CUSTOMERS} \\
\hline
\mathit{OrdinaryCust}' \cap \mathit{PriorityCust}' = \varnothing \\
\mathit{OrdinaryCust}' \cup \mathit{PriorityCust}' = \mathit{RegisteredCust}' \\
\hline
\end{array}$$

These schemas can then be combined to form a schema which describe the invariant properties of the system:

$$\begin{array}{|l}
\Delta \mathit{CustSys} \\
\hline
\mathit{cust} \\
\mathit{cust}' \\
\hline
\end{array}$$

This is equivalent to the schema

$$\begin{array}{|l}
\Delta \mathit{CustSys} \\
\hline
\mathit{OrdinaryCust}, \mathit{PriorityCust}, \mathit{RegisteredCust} : \mathbb{F}\ \mathit{CUSTOMERS} \\
\mathit{OrdinaryCust}', \mathit{PriorityCust},' \mathit{RegisteredCust}' : \mathbb{F}\ \mathit{CUSTOMERS} \\
\hline
\mathit{OrdinaryCust} \cap \mathit{PriorityCust} = \varnothing \\
\mathit{OrdinaryCust} \cup \mathit{PriorityCust} = \mathit{RegisteredCust} \\
\mathit{OrdinaryCust}' \cap \mathit{PriorityCust}' = \varnothing \\
\mathit{OrdinaryCust}' \cup \mathit{PriorityCust}' = \mathit{RegisteredCust}' \\
\hline
\end{array}$$

As will be explained in the next section the symbol Δ is used to identify a schema which describes a change to stored data.

Schemas can also be extended by adding to their signatures or predicates. Adding a new declaration to a signature is achieved by following the schema name by a semicolon and the new declaration. For example, if the schema *PermanentFiles* is

$$\begin{array}{|l}
\textit{PermanentFiles} \\
\hline
\textit{PfileDirectory} : \mathbb{F}\ \textit{FILES} \\
\textit{MasterFile} : \textit{FILES} \\
\hline
\textit{MasterFile} \in \textit{PfileDirectory} \\
\hline
\end{array}$$

then the schema

$$\textit{new} \mathrel{\widehat{=}} [\textit{PermanentFiles};\ \textit{UserDirectory} : \mathbb{F}\ \textit{FILES}]$$

would be

$$\begin{array}{|l}
\textit{new} \\
\hline
\textit{UserDirectory}, \textit{PfileDirectory} : \mathbb{F}\ \textit{FILES} \\
\textit{MasterFile} : \textit{FILES} \\
\hline
\textit{MasterFile} \in \textit{PfileDirectory} \\
\hline
\end{array}$$

A predicate can be added to a schema by concatenating the schema's name with the symbol |, followed by the predicate to be added. Thus,

$$\textit{NewPerm} \mathrel{\widehat{=}} [\textit{PermanentFiles} \mid \#\textit{PfileDirectory} \leq 200]$$

is equivalent to the schema

$$\begin{array}{|l}
\textit{NewPerm} \\
\hline
\textit{PfileDirectory} : \mathbb{F}\ \textit{FILES} \\
\textit{MasterFile} : \textit{FILES} \\
\hline
\textit{MasterFile} \in \textit{PfileDirectory} \\
\#\textit{PfileDirectory} \leq 200 \\
\hline
\end{array}$$

10.3 Z schema conventions

When constructing Z specifications, a number of conventions are observed. These conventions dictate the form of schema and variable names and provide a visual cue to the reader about the function and nature of schemas and the variables used in the schemas.

The first convention is concerned with schemas that describe events which change part of a system. The symbol Δ is used to precede a schema name to indicate the general properties of the schema that are unaffected by a series of events. As an example, consider the schema *SpoolQueueSys* which describes the fact that all elements of a spool queue of files awaiting printing are temporary files.

$$\begin{array}{|l}
\hline
SpoolQueueSys \\
\hline
TempFiles : \mathbb{F}\ FILES \\
SpoolQueue : \mathrm{seq}\ FILES \\
\hline
\mathrm{ran}\ SpoolQueue \subseteq TempFiles \\
\hline
\end{array}$$

The schema which describes that this invariant property will always hold is called $\Delta SpoolQueueSys$

$$\begin{array}{|l}
\hline
\Delta SpoolQueueSys \\
\hline
SpoolQueueSys \\
SpoolQueueSys' \\
\hline
\end{array}$$

which is, of course, equivalent to

$$\begin{array}{|l}
\hline
\Delta SpoolQueueSys \\
\hline
TempFiles, TempFiles' : \mathbb{F}\ FILES \\
SpoolQueue, SpoolQueue' : \mathrm{seq}\ FILES \\
\hline
\mathrm{ran}\ SpoolQueue \subseteq TempFiles \\
\mathrm{ran}\ SpoolQueue' \subseteq TempFiles' \\
\hline
\end{array}$$

It can be used in any schema which does not affect the fact that all the files in a spool queue are temporary files. For example, the schema *RemoveSpoolFile* describes the event of removing the first element of the spool queue.

$$\begin{array}{|l}
\hline
RemoveSpoolFile \\
\hline
\Delta SpoolQueueSys \\
RemovedFile! : FILES \\
\hline
RemovedFile! = head\ SpoolQueue \\
SpoolQueue' = tail\ SpoolQueue \\
\hline
\end{array}$$

Worked example 10.1 A computerized library system at a university administers the loan of books to staff and students alike. Each user has a unique borrower number and each book in the library has a unique reference number. Staff can borrow no more than 20 books, while students borrow no more than 10 books. There are no other users apart from staff and students. Write down the schema $\Delta users$ which describes the invariant properties of the library system. Assume that the fact that a user borrows a series of books is modelled by means of a relation over user names and book registrations.

Solution The invariant properties of the system are:

- The only users are staff and students.
- A user cannot simultaneously be a member of staff and a student.
- No member of staff can borrow more than 20 books.
- No student can borrow more than 10 books.
- A book can only be borrowed by one user.

These properties are embodied in the schema *users*:

$$
\begin{array}{l}
\textit{users} \\
\hline
\textit{staff}, \textit{students}, \textit{AllUsers} : \mathbb{F}\ \textit{BNUMBERS} \\
\textit{borrows} : \textit{BNUMBERS} \rightarrowtail\!\!\!\!\!\rightarrow \mathbb{F}\ \textit{BREFNOS} \\
\hline
\textit{staff} \cup \textit{students} = \textit{AllUsers} \\
\textit{staff} \cap \textit{students} = \varnothing \\
\forall\, st : \textit{staff} \bullet \\
\qquad \#(\textit{borrows}\ st) \leq 20 \\
\forall\, stud : \textit{students} \bullet \\
\qquad \#(\textit{borrows}\ stud) \leq 10 \\
\forall\, x, y : \operatorname{ran} \textit{borrows} \bullet \\
\qquad x \neq y \Rightarrow x \cap y = \varnothing \\
\hline
\end{array}
$$

The invariant properties can then be expressed as

$$
\begin{array}{l}
\Delta \textit{users} \\
\hline
\textit{users} \\
\textit{users}' \\
\hline
\end{array}
$$

■

Worked example 10.2 A system for monitoring the operation of a chemical plant is to be developed by a software house. The system will consist of a series of monitors (thermocouples, pressure transducers) each of which will be attached to a chemical reactor. Each monitor is also connected to a single computer. A monitor is identified by a unique monitor name; each computer is identified by a unique computer name. There are two types of computers which will be used in the system. Type A computers are very powerful and are able to process signals from up to 200 monitors. Type B computers are less powerful, older computers which can process signals from up to 120 monitors. The fact that a computer is connected to monitors is modelled by a function from computer names to finite sets of monitor names. The fact that a monitor is attached to a reactor is modelled by means of a partial injective function which maps a monitor name to a monitoring point on a reactor. Write down a schema which describes the invariant properties of the proposed system.

Solution The invariant properties are:

- No computer can be both type A and type B.
- No type A computer can be connected to more than 200 monitoring points.
- No type B computer can be connected to more than 120 monitoring points.
- If ta is the number of type A computers in the system and tb is the number of type B computers in the system then $ta * 200 + tb * 120$ must be greater than or equal to the number of connections to the reactors.
- No monitor is connected to more than one computer.

This can be expressed in the schema

$$
\begin{array}{|l}
\hline
\Delta MonSys \\
MonSys \\
MonSys' \\
\hline
\end{array}
$$

where $MonSys$ is

$$
\begin{array}{|l}
\hline
\textit{MonSys} \\
\textit{AllComputers}, \textit{TypeAs}, \textit{TypeBs} : \mathbb{F}\ \textit{COMPNAMES} \\
\textit{HasMonitor} : \textit{COMPNAMES} \twoheadrightarrow \mathbb{F}\ \textit{MONNAMES} \\
\textit{IsConnected} : \textit{MONNAMES} \rightarrowtail\!\!\!\!\!\!\twoheadrightarrow \textit{MONPOINTS} \\
\hline
\textit{AllComputers} = \textit{TypeAs} \cup \textit{TypeBs} \\
\operatorname{dom} \textit{IsConnected} = \bigcup \{\ c : \operatorname{dom} \textit{HasMonitor} \bullet \textit{HasMonitor}\ c\ \} \\
\textit{TypeAs} \cap \textit{TypeBs} = \varnothing \\
\forall\ \textit{comp} : \textit{TypeAs} \bullet \\
\quad \#(\textit{HasMonitor}\ \textit{comp}) \leq 200 \\
\forall\ \textit{comp} : \textit{TypeBs} \bullet \\
\quad \#(\textit{HasMonitor}\ \textit{comp}) \leq 120 \\
\#\textit{TypeAs} * 200 + \#\textit{TypeBs} * 120 \geq \#\operatorname{dom} \textit{IsConnected} \\
\forall\ \textit{co1}, \textit{co2} : \operatorname{ran} \textit{HasMonitor} \bullet \\
\quad \textit{co1} \neq \textit{co2} \Rightarrow \textit{co1} \cap \textit{co2} = \varnothing \\
\hline
\end{array}
$$

■

A further convention involves schemas which describe objects or sets of objects which are not changed by events. Such schemas are named by preceding a schema name with the symbol Ξ. Δ schemas are those which indicate invariant properties, while Ξ schemas are used to describe parts of a system that do not change.

For example, consider the commercial data processing system described by the signature

$$
\begin{array}{l}
\textit{ProductPrice} : \textit{ProductNos} \twoheadrightarrow \textit{prices} \\
\textit{ProductSupplier} : \textit{suppliers} \leftrightarrow \textit{ProductNos} \\
\textit{ProductStock} : \textit{ProductNos} \twoheadrightarrow \mathbb{N} \\
\textit{OrderLevel} : \textit{ProductNos} \twoheadrightarrow \mathbb{N} \\
\textit{ProductPosition} : \textit{ProductNos} \twoheadrightarrow \textit{WarehousePositions} \\
\textit{OrderProduct} : \textit{OrderNos} \twoheadrightarrow \textit{ProductNos} \\
\textit{BackOrders} : \textit{ProductNos} \twoheadrightarrow \operatorname{seq} \textit{OrderNos} \\
\textit{OrderPlaced} : \textit{OrderNos} \twoheadrightarrow \textit{CustomerNos}
\end{array}
$$

inside a schema *DPOrder*. *ProductPrice* models the association between a product and its price; *ProductSupplier* models the association between a supplier and a series of products that are supplied; *ProductStock* models the association between a product and the minimum stock level below which it must never be allowed to drop. *ProductPosition* models the association between a product and its position in a warehouse. It is used to generate picking lists. Such lists give the names of each product to be removed by warehouse staff in response to an order. They are arranged in ascending order of warehouse position. This enables warehouse staff to travel the warehouse in one smooth journey, rather than travelling large distances from one end of the warehouse to the other. *OrderLevel* gives the current number of items of a product currently in stock. *OrderProduct* models the association between an order number and a product being ordered, *BackOrders* is used to model the queues of orders for products which are not in stock, and *OrderPlaced* models the association between an order and the customer that placed the order.

Some events in such a system will not affect all the variables in the signature shown. For example, part of the system may be concerned with processing queries from order clerks. For such events the schema $\Xi DPOrder$ can be defined as

$$
\begin{array}{|l}
\hline \Xi DPOrder \\
DPOrder \\
DPOrder' \\
\hline
PredicatePart \\
\hline
\end{array}
$$

where *PredicatePart* will contain the fact that each component of the signature was equal to its new value. It could, for example, contain statements such as

$$OrderProduct' = OrderProduct$$
$$BackOrders' = BackOrders$$

It could also contain the operator θ. This returns the variables of the signature of a schema. Thus, $\Xi DPOrder$ would be written more conveniently as

$$
\begin{array}{|l}
\hline \Xi DPOrder \\
DPOrder \\
DPOrder' \\
\hline
\theta DPOrder' = \theta DPOrder \\
\hline
\end{array}
$$

As an example of the use of the Ξ convention, consider the schema which represents the event that occurs when an order clerk queries the system to obtain the number of *BackOrders* for a particular product. This can be written as

BackOrderQuery

$\Xi DPOrder$
$ProductNumber? : ProductNos$
$number! : \mathbb{N}$

$ProductNumber? \in \mathrm{dom}\ BackOrders$

$number! = \#BackOrders\ ProductNumber?$

A final convention concerns the variables in a signature. Some variables in the signature of a schema will represent input variables whose values are to be used by the event described by the schema. Other variables will represent output variables which will gain a value after the event described by the schema has occurred. For example, the schema

CustQuery

$\Xi updates$
$CustomerNo? : \mathbb{N}$
$AccountBalance! : \mathbb{N}$

$CustomerNo? \in \mathrm{dom}\ balance$

$AccountBalance! = balance\ CustomerNo$

describes a query in a banking system where *balance* is a partial function which maps customer accounts into the balance of cash in that account. The query provides a customer account number *CustomerNo*? and the balance is returned in *AccountBalance*!, In the schema *CustomerNo*? is an input variable while *AccountBalance*! is an output variable. In Z schemas the input variables are annotated with question marks: output variables are annotated by exclamation marks.

Exercise 10.2

A banking system is to be specified using Z. Each bank in the system contains a series of current accounts and deposit accounts. An account is identified by a unique number. An account is owned by a customer who is also uniquely identified by a customer name. A customer can own a number of accounts.

(i) Write down a signature which describes the banking system.
(ii) Write down the schema $\Delta BankSys$.
(iii) Write down the schema $\Xi BankSys$.
(iv) Write down the schema *DepCustOne* which describes the event that occurs when those customers who have at least one deposit account are listed.
(v) Write down the schema *CurCustomersOne* which describes the event that occurs when those customers who have at least one current

account are listed.

(vi) Write down the schema *BothAccounts* which describes the event that occurs when those customers who have at least one current account and at least one deposit account are listed.

(vii) Write down a schema *balance* which describes the event that occurs when the balance of a particular account is displayed.

■

10.4 Logical operators and schemas

Schemas can be combined using logical operators. The effect of applying a binary logical operator to two schemas is a schema whose signature is the union of the signature of the operands' signatures provided that the types of shared variables agree. The predicate of the new schema so formed is constructed by applying the logical operator to the predicates of the schemas which are its operands.

For example, if the schema *accounts* is

```
┌─ accounts ───────────────────────────────────────
│ HasAccount : CustomerNames ⇸ 𝔽 AccountNos
│ LargeHolders : 𝔽 CUSTNAMES
├──────────────
│ ∀ name : LargeHolders •
│     #(HasAccount name) > 10
└──────────────────────────────────────────────────
```

and if the schema *ComAccounts* is

```
┌─ ComAccounts ────────────────────────────────────
│ CompanyAccounts : 𝔽 CUSTNAMES
│ LargeHolders : 𝔽 CUSTNAMES
├──────────────
│ CompanyAccounts ⊆ LargeHolders
└──────────────────────────────────────────────────
```

then the schema

$$AllAccounts \mathrel{\widehat=} accounts \wedge ComAccounts$$

is

$$
\begin{array}{|l}
\multicolumn{1}{l}{\underline{\textit{AllAccounts}\qquad\qquad\qquad\qquad\qquad\qquad\qquad\qquad}} \\
\textit{CompanyAccounts} : \mathbb{F}\ \textit{CUSTNAMES} \\
\textit{LargeHolders} : \mathbb{F}\ \textit{CUSTNAMES} \\
\textit{HasAccount} : \textit{CustomerNames} \mathrel{+\!\!\!\!\to} \mathbb{F}\ \textit{AccountNos} \\
\hline
\textit{CompanyAccounts} \subseteq \textit{LargeHolders} \\
\forall\, \textit{name} : \textit{LargeHolders} \bullet \\
\qquad \#(\textit{HasAccount name}) > 10 \\
\hline
\end{array}
$$

10.5 Using the schema calculus—an example

The Reliable Parts Company provide spare parts for garages in a major city. The parts are identified by a unique name. Many of the parts can only be fitted to one type of vehicle. However, some parts are able to be fitted to a range of vehicles. The company also stocks a number of parts which are supplied by different manufacturers and which are equivalent in the sense that they can be fitted to the same type of car.

The spare parts data base for this application can be modelled by the signature

$$
\begin{array}{l}
\textit{VehiclesSupplied} : \mathbb{F}\ \textit{VEHICLES} \\
\textit{SuppOfPart} : \textit{PARTNOS} \mathrel{+\!\!\!\!\to} \mathbb{F}\ \textit{SUPPLIERS} \\
\textit{PriceOfPart} : \textit{PARTNOS} \mathrel{+\!\!\!\!\to} \textit{PartPrices} \\
\textit{InStock} : \textit{PARTNOS} \mathrel{+\!\!\!\!\to} \mathbb{N} \\
\textit{FittedTo} : \textit{PARTNOS} \mathrel{+\!\!\!\!\to} \mathbb{F}\ \textit{VEHICLES} \\
\textit{substitute} : \textit{PARTNOS} \mathrel{+\!\!\!\!\to} \mathbb{F}\ \textit{PARTNOS}
\end{array}
$$

VehiclesSupplied is the set of all vehicles that the company supplies spare parts for. The relation *SuppOfPart* models the association between a part number and the suppliers who supply the part. *PriceOfPart* models the association between a part and its price. *InStock* models the association between a part and the number in stock of the part. *FittedTo* models the association between a part and the vehicles it can be fitted to. Finally *substitute* models the association between a part and all the parts that can be substituted for it. The part of the statement of requirements concerned with queries follows.

> 4. A number of queries can be typed in by a stores clerk.
>
> 4.1 LISTINSTOCK. A stores clerk can type in the LISTINSTOCK command at a terminal and the system will respond by displaying all the part numbers and quantities in stock of all current parts.
>
> 4.2 INSTOCK. A stores clerk can type in the INSTOCK command at a terminal together with a part number. The system will respond

by displaying the quantity in stock of the particular part whose number is typed, together with the number and quantity in stock of any substitute parts.

4.3 PRICEPARTS. A stores clerk can type in the PRICEPARTS command at a terminal together with a part number. The system will respond by displaying the price of the part whose number has been typed, together with the number and price of any substitutes.

4.4 SUPPLIERPARTS. A stores clerk can type in the SUPPLIERPARTS command at a terminal together with a part number. The system will respond by displaying the names of the suppliers of the part whose number has been typed.

4.5 PRINTCARLIST. A stores clerk can type in the PRINTCARLIST command at a terminal together with a vehicle name. The system will respond by printing the names of those parts which can be fitted to the vehicle.

The schema used to describe properties of the system will be called *DbProp*.

$$
\begin{array}{l}
\textit{DbProp} \\
\hline
\textit{VehiclesSupplied} : \mathbb{F}\ \textit{VEHICLES} \\
\textit{SuppOfPart} : \textit{PARTNOS} \twoheadrightarrow \mathbb{F}\ \textit{SUPPLIERS} \\
\textit{PriceOfPart} : \textit{PARTNOS} \twoheadrightarrow \textit{PartPrices} \\
\textit{InStock} : \textit{PARTNOS} \twoheadrightarrow \mathbb{N} \\
\textit{FittedTo} : \textit{PARTNOS} \twoheadrightarrow \mathbb{F}\ \textit{VEHICLES} \\
\textit{substitute} : \textit{PARTNOS} \twoheadrightarrow \mathbb{F}\ \textit{PARTNOS} \\
\hline
\operatorname{dom} \textit{SuppOfPart} = \operatorname{dom} \textit{PriceOfPart} \\
\operatorname{dom} \textit{SuppOfPart} = \operatorname{dom} \textit{InStock} \\
\operatorname{dom} \textit{SuppOfPart} = \operatorname{dom} \textit{FittedTo} \\
\operatorname{dom} \textit{substitute} \subseteq \operatorname{dom} \textit{SuppOfPart} \\
\bigcup(\operatorname{ran} \textit{substitute}) \subseteq \operatorname{dom} \textit{SuppOfPart} \\
\forall p : \operatorname{dom} \textit{substitute} \bullet \\
\quad p \notin \textit{substitute}\ p \\
\hline
\end{array}
$$

The schema $\Delta \textit{DataBase}$ can be expressed in terms of *DbProp* as

$$\Delta \textit{DataBase} \ \widehat{=}\ \textit{DbProp}' \wedge \textit{DbProp}$$

The schema which can be used in queries and which describes the fact that the data base is unchanged is $\Xi \textit{DataBase}$.

$$\begin{array}{|l}
\multicolumn{1}{l}{\Xi DataBase} \\
\hline
\Delta DataBase \\
\hline
\theta DbProp' = \theta Dbprop \\
\hline
\end{array}$$

The schema *ListInStock* describes the event that occurs when the LISTIN-STOCK command is invoked.

$$\begin{array}{|l}
\multicolumn{1}{l}{ListInStock} \\
\hline
\Xi DataBase \\
NamesAndQuant! : PARTNOS \twoheadrightarrow \mathbb{N} \\
\hline
NamesAndQuant! = InStock \\
\hline
\end{array}$$

The schema *InStock* describes the event that occurs when the INSTOCK command is typed.

$$\begin{array}{|l}
\multicolumn{1}{l}{InStock} \\
\hline
\Xi DataBase \\
TypedNo? : PARTNOS \\
stocks! : PARTNOS \twoheadrightarrow \mathbb{N} \\
\hline
TypedNo? \in \operatorname{dom} InStock \\
stocks! = (\{TypedNo?\} \cup (substitute\ TypedNo?)) \lhd InStock \\
\hline
\end{array}$$

The schema *PriceParts* describes the event that occurs when the PRI-CEPARTS command is typed.

$$\begin{array}{|l}
\multicolumn{1}{l}{PriceParts} \\
\hline
\Xi DataBase \\
TypedNo? : PARTNOS \\
prices! : PARTNOS \twoheadrightarrow PartPrices \\
\hline
TypedNo? \in \operatorname{dom} InStock \\
prices! = (\{TypedNo?\} \cup (substitute\ TypedNo?)) \lhd PriceOfPart \\
\hline
\end{array}$$

The schema *SupplierParts* describes the event that occurs when the SUP-PLIERPARTS command is typed.

$$\begin{array}{|l}
\multicolumn{1}{l}{_\ SupplierParts\ ________________________________} \\
\Xi DataBase \\
suppliers! : \mathbb{F}\ SUPPLIERS \\
TypedNo? : PARTNOS \\
\hline
TypedNo? \in \mathrm{dom}\ InStock \\
suppliers! = SuppOfPart\ TypedNo? \\
\hline
\end{array}$$

Finally, the schema *PrintCarList* describes the event that occurs when the PRINTCARLIST command is typed.

$$\begin{array}{|l}
\multicolumn{1}{l}{_\ PrintCarList\ ________________________________} \\
\Xi DataBase \\
TypedVehicle? : VEHICLES \\
PartsFitted! : \mathbb{F}\ PARTNOS \\
\hline
TypedVehicle? \in VehiclesSupplied \\
PartsFitted! = \{\ p : \mathrm{dom}\ InStock \mid TypedVehicle? \in FittedTo\ p\ \} \\
\hline
\end{array}$$

Success or failure can be associated with a schema *CommandCondition*

$$\begin{array}{|l}
\multicolumn{1}{l}{_\ CommandCondition\ ____________________________} \\
report! : conditions \\
\hline
\end{array}$$

where *report*! gives an indication of the success and failure of an event. In the schemas described above there can be two types of event. First, the number of a part typed in by a clerk would not refer to a part whose details are stored in the data base. Second a vehicle name is typed which is not supplied by the company. These errors can be defined by the schemas *PartError* and *VehicleError* which make use of elements of the set *conditions* defined as

$$conditions == \{InvalidPartNo, InvalidVehicleName, successful\}$$

The schemas are shown below.

$$\begin{array}{|l}
\multicolumn{1}{l}{_\ PartError\ ____________________________________} \\
\Xi DataBase \\
CommandConditions \\
TypedNo? : PARTNOS \\
\hline
TypedNo? \notin \mathrm{dom}\ InStock \\
report! = InvalidPartNo \\
\hline
\end{array}$$

$$
\begin{array}{|l}
\hline
\textit{VehicleError} \\
\Xi DataBase \\
CommandConditions \\
TypedVehicle? : VEHICLES \\
\hline
TypedVehicle \notin VehiclesSupplied \\
report! = InvalidVehicleName \\
\hline
\end{array}
$$

Finally success can be defined as

$$
\begin{array}{|l}
\hline
\textit{success} \\
CommandConditions \\
\hline
report! = successful \\
\hline
\end{array}
$$

The full effect of the commands can now be defined using the logical operators $\wedge$ and $\vee$.

$$
\begin{array}{l}
FullListInStock \;\widehat{=}\; (ListInStock \wedge success) \\
FullInStock \;\widehat{=}\; (InStock \wedge success) \vee PartError \\
FullPriceParts \;\widehat{=}\; (PriceParts \wedge success) \vee PartError \\
FullSupplierParts \;\widehat{=}\; (SupplierParts \wedge success) \vee PartError \\
FullPrintCarList \;\widehat{=}\; (PrintCarList \wedge success) \vee VehicleError
\end{array}
$$

For example, *FullInStock* is then equivalent to the complete schema shown as Figure 10.1

$$
\begin{array}{|l}
\hline
\textit{FullInStock} \\
\textit{VehiclesSupplied}', \textit{VehiclesSupplied} : \mathbb{F}\ \textit{VEHICLES} \\
\textit{SuppOfPart}, \textit{SuppOfPart}' : \textit{PARTNOS} \nrightarrow \mathbb{F}\ \textit{SUPPLIERS} \\
\textit{PriceOfPart}, \textit{PriceOfPart}' : \textit{PARTNOS} \nrightarrow \textit{PartPrices} \\
\textit{InStock}, \textit{InStock}' : \textit{PARTNOS} \nrightarrow \mathbb{N} \\
\textit{FittedTo}, \textit{FittedTo}' : \textit{PARTNOS} \nrightarrow \mathbb{F}\ \textit{VEHICLES} \\
\textit{substitute}, \textit{substitute}' : \textit{PARTNOS} \nrightarrow \mathbb{F}\ \textit{PARTNOS} \\
\textit{TypedNo}? : \textit{PARTNOS} \\
\textit{stocks}! : \textit{PARTNOS} \nrightarrow \mathbb{N} \\
\textit{report}! : \textit{conditions} \\
\hline
\mathrm{dom}\, \textit{SuppOfPart} = \mathrm{dom}\, \textit{PriceOfPart} \\
\mathrm{dom}\, \textit{SuppOfPart} = \mathrm{dom}\, \textit{InStock} \\
\mathrm{dom}\, \textit{SuppOfPart} = \mathrm{dom}\, \textit{FittedTo} \\
\mathrm{dom}\, \textit{substitute} \subseteq \mathrm{dom}\, \textit{SuppOfPart} \\
\bigcup(\mathrm{ran}\, \textit{substitute}) \subseteq \mathrm{dom}\, \textit{SuppOfPart} \\
\forall p : \mathrm{dom}\, \textit{substitute} \bullet \\
\quad p \notin \textit{substitute}\ p \\
\mathrm{dom}\, \textit{SuppOfPart}' = \mathrm{dom}\, \textit{PriceOfPart}' \\
\mathrm{dom}\, \textit{SuppOfPart}' = \mathrm{dom}\, \textit{InStock}' \\
\mathrm{dom}\, \textit{SuppOfPart}' = \mathrm{dom}\, \textit{FittedTo}' \\
\mathrm{dom}\, \textit{substitute}' \subseteq \mathrm{dom}\, \textit{SuppOfPart}' \\
\bigcup(\mathrm{ran}\, \textit{substitute}') \subseteq \mathrm{dom}\, \textit{SuppOfPart}' \\
\forall p : \mathrm{dom}\, \textit{substitute}' \bullet \\
\quad p \notin \textit{substitute}'\ p \\
\textit{VehiclesSupplied}' = \textit{VehiclesSupplied} \\
\textit{SuppOfPart}' = \textit{SuppOfPart} \\
\textit{PriceOfPart}' = \textit{PriceOfPart} \\
\textit{InStock}' = \textit{InStock} \\
\textit{FittedTo}' = \textit{FittedTo} \\
\textit{substitute}' = \textit{substitute} \\
(\textit{TypedNo}? \in \mathrm{dom}\, \textit{InStock} \land \\
\quad \textit{stocks}! = (\{\textit{TypedNo}?\} \cup (\textit{substitute}\ \textit{TypedNo}?)) \lhd \textit{InStock} \\
\quad \textit{report}! = \textit{success}) \\
\lor (\textit{TypedNo}? \notin \mathrm{dom}\, \textit{InStock} \land \\
\quad \textit{report}! = \textit{InvalidPartNo}) \\
\hline
\end{array}
$$

Figure 10.1 The schema *FullInStock*

11

Some small Z examples

Aims

- To describe some small but relatively realistic examples of Z specifications.

This section describes three examples of Z specifications. Each specification describes a part of a relatively realistic software system, with some degree of simplification introduced in order to keep the chapter to a manageable size.

11.1 A block handler

One of the more important parts of a computer's operating system is the subsystem which maintains files which have been created by users. Part of the filing subsystem is the block handler. Files in the file store will be made up of blocks of storage which are held on some file storage device. During the operation of the computer, files will be created and deleted. This means that, in the first case, blocks will be needed to form a file and, in the second case, blocks will be released from a file ready for use in another file. In order to cope with this the filing subsystem will maintain a reservoir of unused blocks, and will also keep track of blocks which are currently in use. When blocks are released from a deleted file they are normally added to a queue of blocks waiting to be added to the reservoir of unused blocks. This is shown in Figure 11.1. In this figure a number of components are shown: the reservoir of free blocks, the blocks which currently make up the files administered by the operating system, and those blocks which are waiting to be added to the reservoir. The waiting blocks are held in a queue, with each element of the queue containing a set of blocks from a deleted file. The invariant which describes the block handler has a number of conditions:

- No block will be marked as both free and used.

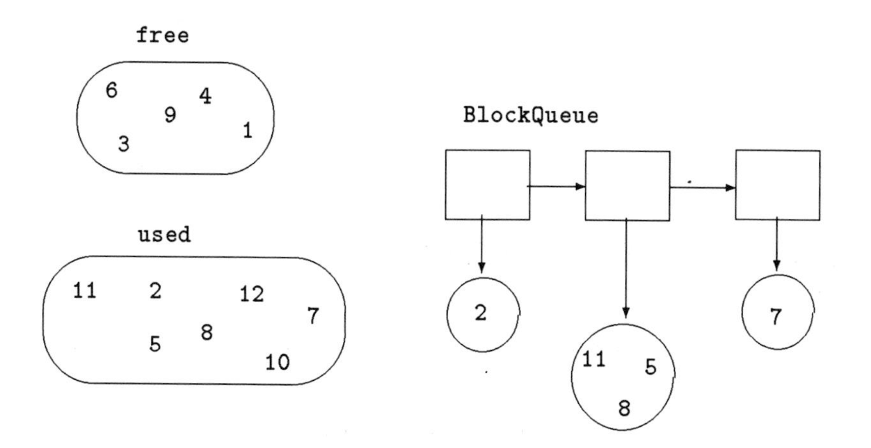

Figure 11.1 A block handler

- All the sets of blocks held in the queue will be subsets of the collection of currently used blocks.

- There will be no elements of the queue which will contain the same block numbers.

- The collection of used blocks and blocks that are free will be the total collection of blocks that make up files.

The types used in the block handler are shown below. *MaxBlocks* is the largest block number available.

$[BLOCKS]$

$AllBlocks = 1 \mathrel{..} MaxBlocks$

The schemas which describes the data invariant are shown below.

$$
\begin{array}{|l}
\hline
\Delta BlockHandler \\
BlockHandler \\
BlockHandler' \\
\hline
\end{array}
$$

$$
\begin{array}{|l}
\multicolumn{1}{l}{\_\_\mathit{BlockHandler}\_\_\_\_\_\_\_\_\_\_\_\_\_\_\_\_\_\_\_\_}\\
\mathit{used}, \mathit{free} : \mathbb{F}\ \mathit{BLOCKS}\\
\mathit{BlockQueue} : \mathrm{seq}(\mathbb{F}\ \mathit{BLOCKS})\\
\hline
\mathit{used} \cap \mathit{free} = \varnothing\\
\forall i : \mathrm{dom}\ \mathit{BlockQueue} \bullet\\
\quad \mathit{BlockQueue}\ i \subseteq \mathit{used}\\
\forall i, j : \mathrm{dom}\ \mathit{BlockQueue} \bullet\\
\quad i \neq j \Rightarrow \mathit{BlockQueue}\ i \cap \mathit{BlockQueue}\ j = \varnothing\\
\mathit{used} \cup \mathit{free} = \mathit{AllBlocks}\\
\hline
\end{array}
$$

Let us assume that there are two operations that are required. The first is to remove a set of blocks which is at the head of the queue of returned blocks. The schema which describes this is shown below.

$$
\begin{array}{|l}
\multicolumn{1}{l}{\_\_\mathit{RemoveBlocks}\_\_\_\_\_\_\_\_\_\_\_\_\_\_\_\_\_\_\_\_}\\
\Delta \mathit{BlockHandler}\\
\hline
\#\mathit{BlockQueue} > 0\\
\mathit{used}' = \mathit{used} \setminus \mathit{head}\ \mathit{BlockQueue}\\
\mathit{free}' = \mathit{free} \cup \mathit{head}\ \mathit{BlockQueue}\\
\mathit{BlockQueue}' = \mathit{tail}\ \mathit{BlockQueue}\\
\hline
\end{array}
$$

The first predicate is the pre-condition which states that the queue must have at least one element; the second predicate describes the removal of the returned blocks from the collection of used blocks; the third predicate describes the addition of the returned blocks to the collection of free blocks; and the fourth predicate describes the removal of the head of the queue.

Worked example 11.1 Write down a schema which describes the addition of a set of blocks $ABlocks?$ to the queue of blocks awaiting assignment to the reservoir of free blocks.

Solution The schema is shown below.

$$
\begin{array}{|l}
\multicolumn{1}{l}{\_\_\mathit{AddBlocks}\_\_\_\_\_\_\_\_\_\_\_\_\_\_\_\_\_\_\_\_}\\
\Delta \mathit{BlockHandler}\\
\mathit{ABlocks?} : \mathbb{F}\ \mathit{BLOCKS}\\
\hline
\mathit{ABlocks?} \subseteq \mathit{used}\\
\mathit{BlockQueue}' = \mathit{BlockQueue} \frown \langle \mathit{ABlocks?} \rangle\\
\hline
\end{array}
$$

The first predicate is the pre-condition which states that the blocks to be added should be marked as used and the second predicate describes the addition of the blocks.

■

11.2 An identifier table

The second example is that of an identifier table. These tables are used extensively in system software where they hold sequences of characters (identifiers) which describe some entity such as the name of a program variable, the name of a process, or the name of a file. An identifier table has the property that its elements are unique, together with the fact that it usually has some upper limit on the number of identifiers which can be stored. This example involves a table of identifiers which can contain a maximum of *MaxIds* identifiers. The state can be described as

$[IDENTIFIERS]$

SymTable

$table : \mathbb{F}\ IDENTIFIERS$

$\#table \leq MaxIds$

The schema which characterizes the data invariant is

$\Delta SymTable$

$SymTable$
$SymTable'$

and the schema which describes the fact that the symbol table will be unaffected by an operation is

$\Xi SymTable$

$\Delta SymTable$

$\theta table' = \theta table$

Three operations are required: an operation which adds an identifier to the table, an operation which removes an identifier from a table, and an operation which checks that an identifier is in the table.

The first operation is shown below:

AddId

$id? : IDENTIFIERS$
$\Delta SymTable$

$\#table < MaxIds$

$id? \notin table$

$table' = table \cup \{id?\}$

The first two predicates represent the pre-condition: that the table must hold less than *MaxIds* identifiers and that the identifier to be added must not be in the table. The final predicate describes the addition of the identifier to the table.

Worked example 11.2 Write down a schema which removes an identifier *id?* from a table.

Solution The schema is shown below.

RemoveId

$id? : IDENTIFIERS$
$\Delta SymTable$

$id? \in table$

$table' = table \setminus \{id?\}$

The first predicate is the pre-condition which describes the fact that the identifier should be in the table, the second predicate describes the removal of the identifier.
■

The final operation is one which checks whether an identifier is contained in the table. For this a type definition is needed. The individual values of the type represent the successful or unsuccessful search for an identifier. This is shown below:

$$results ::= successful \mid unsuccessful$$

The specification of the operation is described by the schema *Findid*.

$$
\begin{array}{|l}
\hline
\textit{FindId} \\
\Xi SymTable \\
id? : IDENTIFIERS \\
found! : results \\
\hline
id? \in table \Rightarrow found! = successful \\
id? \notin table \Rightarrow found! = unsuccessful \\
\hline
\end{array}
$$

11.3 A print spooler

The final example is again taken from the area of system software. It describes a print spooler. In a mainframe or minicomputer operating system a number of users will be instigating print instructions, for example, asking for the contents of a print file to be printed out on a laser printer. Unfortunately, printing devices are slow and the various print requests are usually placed in a queue ready to be printed, with the items to be printed held in a file. When a print request has been satisfied a piece of system software removes the printed file from the queue and issues an instruction to print the next file on the queue.

In a large operating system there will be a number of printers and each one will have a queue associated with it. Moreover, some of the printers can only be economically used for print requests over a particular size.

The state used by a print spooler is illustrated in Figure 11.2. It shows a series of printers each with a unique name. These printers are associated with a queue of files, each of which has a size associated with it. This size represents the number of lines of paper to be generated. Also, as part of the state there is a table which, for each printer, specifies the maximum number of lines allowed for printing requests. I shall assume that one condition imposed on the specification is that no more than $MaxFiles$ are to be found in a queue. The data invariant for this system consists of a number of components:

- There will be no more than $MaxFiles$ in each queue.
- That maximum print line details will be kept for each printer that has a queue.
- That each file name in a queue will not be found in another queue.
- That each file in a queue will be associated with a figure that describes the number of lines to be printed.
- That each file in a queue will contain no more than the maximum allowable lines of printing for the printer with which the queue is associated.

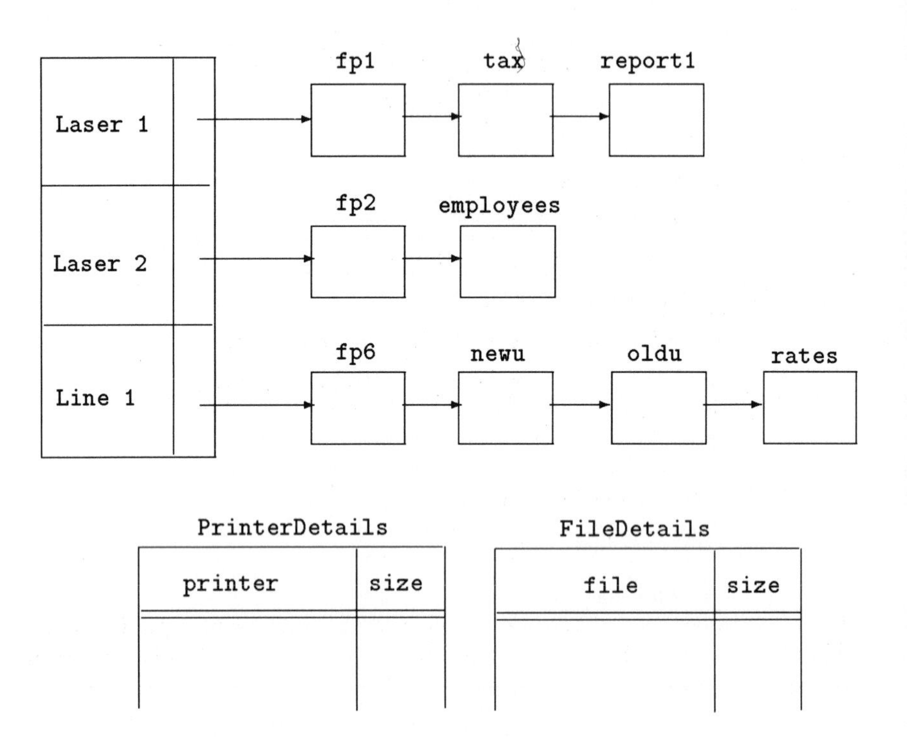

Figure 11.2 The state of a print spooler

The state of the print spooler can be described by the schema *PrintSpooler* shown below. Also shown are the schemas which describe changes to the state and the fact that the state remains unchanged.

$[PRINTERS, FILES]$

$\Delta spooler$
$spooler$
$spooler'$

$\Xi spooler$
$\Delta spooler$

$\theta spooler' = \theta spooler$

spooler

$$queues : PRINTERS \rightarrowtail\!\!\!\!\!\rightarrow \text{seq } FILES$$
$$PrinterDetails : PRINTERS \twoheadrightarrow\!\!\!\!\!\!\!\shortmid\;\; \mathbb{N}$$
$$FileDetails : FILES \twoheadrightarrow\!\!\!\!\!\!\!\shortmid\;\; \mathbb{N}$$

$$\forall i : \text{dom } queues \bullet \\ \quad \# queues\ i \leq MaxFiles$$

$$\text{dom } queues = \text{dom } PrinterDetails$$

$$\forall i, j : \text{dom } queues \bullet \\ \quad i \neq j \Rightarrow \text{ran}(queues\ i) \cap \text{ran}(queues\ j) = \varnothing$$

$$\text{dom } FileDetails = \{ f : FILES \mid \exists i : \text{dom } queues \bullet f \in \text{ran}(queues\ i) \}$$

$$\forall i : \text{dom } queues \bullet \\ \quad \forall j : \text{ran } queues\ i \bullet \\ \quad\quad FileDetails\ j \leq PrinterDetails\ i$$

Let us assume that a number of operations are required. The first is one which interrogates the system to discover how many files are awaiting printing on the printer *print?*

Query

$$\Xi spooler$$
$$print? : PRINTERS$$
$$NoQueued! : \mathbb{N}$$

$$print? \in \text{dom } queues$$

$$NoQueued! = \# queues\ print?$$

The first predicate is the pre-condition which states that the spooler must have details concerning the printer; the second predicate describes the retrieval of the number of items in the printer queue.

The next schema adds a new file *AddFile?* to the queue associated with the printer *prt?* The file *AddFile?* has a length *size?*. The first two predicates in the schema make up the pre-condition. This states that the file to be added is not known to the spooler and the addition of the file to the queue associated with *prt?* does not make the queue size exceed *MaxFiles*. The final three predicates describe the addition of the file to the required queue and the updating of the part of the spooler which keeps details of the print sizes.

AddFile

$\Delta spooler$
$AddFile? : FILES$
$prt? : PRINTERS$
$size? : \mathbb{N}$

$AddFile? \notin \mathrm{dom}\, FileDetails$

$\# queues\ prt? < MaxFiles$
$size? \leq PrinterDetails\ prt?$
$queues'\ prt? = queues\ prt? \frown \langle AddFile? \rangle$

$FileDetails' = FileDetails \cup \{AddFile? \mapsto size?\}$

$PrintDetails' = PrintDetails$

Worked example 11.3 Write down a schema which describes the process of adjusting the print limit of a printer $prt?$ from its present value to a new value $NewLimit?$

Solution The schema is shown below.

ChangePrintLimit

$\Delta spooler$
$prt? : PRINTERS$
$NewLimit? : \mathbb{N}$

$prt? \in \mathrm{dom}\, PrinterDetails$

$PrinterDetails' = PrinterDetails \oplus \{prt? \mapsto NewLimit?\}$

$queues' = queues$

$FileDetails' = FileDetails$

All this does is to override the entry for the printer $prt?$ with a new entry which represents the new limit.
■

The final operation is one which removes a file *file*? from a queue associated with printer $prt1?$ and places it on a second queue associated with the printer $prt2?$

The pre-condition for this operation is that the queue for $prt1?$ contains *file*?, the printers are known to the spooler and that the new queue is not associated with a printer which has a print limit lower than the number of lines of print in *file*? The schema is shown below.

$$
\begin{array}{|l}
\hline
\textit{Transfer} \\
\Delta spooler \\
prt1?, prt2? : PRINTERS \\
file? : FILES \\
\hline
file? \in \mathrm{ran}\ queues\ prt1? \\
prt1? \in \mathrm{dom}\ PrinterDetails \\
prt2? \in \mathrm{dom}\ PrinterDetails \\
PrinterDetails\ prt2? \geq FileDetails\ file? \\
queues'\ prt2? = queues\ prt2? \frown \langle file? \rangle \\
\exists i : 0 \mathrel{..} \# queues'\ prt1? \bullet \\
\quad queues'prt1? \lhd (1 \mathrel{..} i) \frown file? \frown queues'prt1? \lhd \\
\quad (i + 1 \mathrel{..} \# queues'prt1?) = queues\ prt1? \\
\forall i : \mathrm{dom}\ queues \bullet \\
\quad i \neq prt1? \wedge i \neq prt2? \Rightarrow queues'\ i = queues\ i \\
FileDetails' = FileDetails \\
PrinterDetails' = PrinterDetails \\
\hline
\end{array}
$$

The fifth predicate describes the movement of the file to the new queue, the sixth predicate describes the fact that the order of the queue from which the file has been removed is unchanged, and the seventh predicate describes the fact that apart from the two adjusted queues the remaining queues in the system will be unaffected by the operation.

These, then, are three simple but, I hope, practical examples of the use of Z. The next chapter contains a much more substantial example.

12

A large Z specification

Aims

- To revise many of the concepts described in the previous three chapters.
- To show how a large Z specification might be structured.

The purpose of this chapter is to demonstrate some of the power of the schema calculus in incrementally describing a system and to describe how error processing can be accommodated in a Z specification. The final part of the chapter describes one way of organizing a Z specification and outlines how natural language can be used in conjunction with schemas. The specification described in the chapter is for an automated system for a university library. Only selected parts of the statement of requirements are shown; for example, parts 2.8 to 2.11 dealing with the requirements for microfiche hard copy are omitted.

12.1 The statement of requirements

Background

1.1 *The library.* The newly established University of Lincoln has a library serving a mixed community of students and staff. Currently, the system used to administer the loan of books is manual. The statement of requirements describes the functions of a proposed computer-based system.

1.2 *The users.* There are two types of users of the library: staff and students. Staff users are allowed to borrow up to 20 books; student users are allowed to borrow up to 10 books. Both are registered with the library when they join the university. The library currently keeps the name and address of each user on a card file. The borrowing period for both categories is 30 days.

1.3 *The books.* The library currently contains over 2 000 000 books. Details of each book are kept in a card file. Each card in the file gives the book's title, its author(s), and the classification number (ISBN number).

1.4 *The replacement of the manual system.* The manual system is to be replaced by a computer-based system. The system will consist of a centralized minicomputer with a series of VDUs together with plastic tape readers and a printer. This equipment will be used by library assistants and will be positioned at the library issue desk.

Each book will contain a short detachable length of tape which will contain a unique book registration number. This tape will be used to register the borrowing and returning of books. On registration with the library, users will also be issued with a user registration card which contains the user's name, address, and a unique registration number.

1.5 *The borrowing of books.* Users are allowed to borrow books and to reserve books which are currently out on loan. When a book is reserved the user who requires the book is added to a list of current users awaiting the book. When the book is returned, it is not immediately placed on the shelves if it is reserved, but stored safely; the user at the top of the reservation list for the book is then notified and his or her name removed from the list.

1.6 *Missing books.* Occasionally, the library is notified that a book is missing. Usually this happens when a user cannot find a book on the shelves and no record of the book being borrowed can be discovered. When this happens the name of the books is placed on a missing list.

The functions of the system

The following paragraphs describe the functions of the system. It should be borne in mind that errors will occur during the exercise of these functions; for example, the tape reader occasionally malfunctions producing spurious data. Because of this each function should be associated with comprehensive error checking.

2.1 *The registration of new users.* The system should allow a library assistant to register the name and address of a new user together with the user's registration number. The latter will be extracted from a list of unused numbers periodically produced by the system (see paragraph 2.6). When the name and address of the user have been entered the user has the borrowing rights of either a member of staff or a student. On successful completion of registration the library assistant will give the user a registration card containing his name, address, and user registration number. This will be filled out by hand.

2.2 *The removal of existing users.* The system should be able to remove details of a user. For example, when a student user leaves the university, he or she no longer has any borrowing rights. The leaving user will provide a library assistant with their registration card. The assistant will then type in the user's registration number at a VDU. All details of the user corresponding to the registration number are expunged from the system.

2.3 *Changing an address or name.* The system should allow the library assistant to change the address or name of a user. The user provides his registration card and the library assistant uses the VDU to type his registration number and the new address or new name.

2.4 *Registering a new book.* When a new book is purchased by the library it is first given to a librarian who assigns it an ISBN number and an unused registration number. This registration number is provided by a print-out which is generated by the system (see paragraph 2.5). When this process is complete the book is given to a library assistant who registers the book with the system. The assistant types in at a VDU the registration number, the ISBN number, the title, and the author.

2.5 *Generating unused book registration numbers.* When a book is bought by the library, it is assigned a registration number which has not yet been assigned to any catalogued book. A command is provided which allows a library assistant to produce a print-out of currently unused numbers. The assistant should type in the number of registration numbers required and the system will provide the numbers in ascending order. At most 1000 numbers should be provided by the system at one time. The plan for the university library envisages that there will be no more than a million books in stock. Consequently, the registration numbers should have at most six digits.

2.6 *Generating unused user registration numbers.* When a user registers with the library, he or she will be allocated a user registration number which has not been allocated to any other user. These will be taken from a print-out produced by the system. A command should be provided which allows a library assistant to type the number of unused user registration numbers that are required. The system will respond by printing the numbers in ascending order. No more than 1000 numbers should be provided. It is envisaged that no more than 10 000 users will ever be registered with the library at one time. Consequently, registration numbers should have no more than three digits.

2.7 *Generating microfiche hard copy (ISBN order).* The library keeps details of its current books on a microfiche catalogue. Part of the microfiche contains a list of all currently catalogued books arranged in ascending order of ISBN number. Every month this catalogue is photographed for microfiche reproduction. In order to provide hard copy for this process a command should be provided which would enable a library assistant to print out all the catalogued books in the library in the above order.

⋮

2.12 *Notification of overdue books.* Each working day a list of borrowers who have borrowed books which are overdue is prepared. The library assistant responsible for this will load the remote printer with reminder forms. The system will then print the user name, address, book title, and registration number for every overdue book on each remainder form. The system will then only send out one reminder form for a particular book.

⋮

2.21 *The borrowing of books.* When a book is borrowed the borrower gives the registration card to the library assistant on duty. The assistant then extracts the plastic tape from the book and places it in the tape reader. The user number is then typed in. A number of errors may occur during this operation.

- If the user has borrowed the maximum number of books allowed for that category of user, then an error message should be displayed.
- If the system does not recognize the user, then another error message should be displayed.
- Occasionally, a book is borrowed which the system recognizes as being missing. This usually is a result of the book being misfiled on the library shelves. When this happens the system should display an error message. The library assistant will then notify the system that a missing book has been found (see paragraph 2.30). Unless the book is reserved, it will normally be lent to the borrower.
- If a book is borrowed which is recognized as being already borrowed, then the system should display an error. When this occurs the library assistant should take appropriate action. For example, the books should be kept by the library and the borrower contacted.
- If an attempt is made to borrow a book which is reserved, then it should be put to one side and the system should regard the book as awaiting collection by the user who has reserved it.

2.22 *The return of books.* When a book is returned the tape contained in the book is placed in the tape reader and the system marks the book as being on the shelves. A number of errors may occur with this operation.

- The book may not have been borrowed and may be recognized by the system as being on the shelves. A suitable error message should be displayed.
- The system may not recognize the reader. For example, a reader who has been de-registered may return a book for a friend. A suitable error message should be displayed.

- The book may be on the list of missing books. A suitable error message should be displayed. The library assistant would then inform the system that a missing book has been discovered (paragraph 2.30).

After books are returned they are stored behind the issue desk and periodically returned to the shelves. However, if a returned book is reserved by another user, then the book is put to one side and the system will mark the book as awaiting collection by the user who reserved it.

⋮

2.30 *Notification of found books.* At certain times during the operation of the system a book which was believed missing could be discovered. In order to inform the system about this the library assistant should be able to type in the registration number of the book and the system would then mark that book as being on the shelves. If the book is currently reserved, the system should provide a message that it is reserved and add the book to the day's reserved books.

⋮

2.35 *Notification of the return of reserved books.* At the end of the day a library assistant will load the remote printer with notification forms. These forms will inform each user who has reserved a book that the book has been returned that day and is awaiting collection. Each notification form requires the name and address of the user and the title and registration number of the book.

12.2 The static properties of the system

The main objects manipulated by the library system will be books. The statement of requirements outlines the fact that each book will have a unique reference number, an ISBN number, a title, and an author. The statement of requirements also details the fact that books can either be borrowed, on the shelves of the library, or missing. Details of each book can be modelled by the schema *BookDetails*

$$
\begin{array}{l}
\textit{BookDetails} \\
\hline
\textit{HasISBN} : \textit{BOOKNOS} \twoheadrightarrow \textit{ISBNnos} \\
\textit{HasTitle} : \textit{BOOKNOS} \twoheadrightarrow \textit{titles} \\
\textit{HasAuthor} : \textit{BOOKNOS} \twoheadrightarrow \textit{AuthorNames} \\
\hline
\end{array}
$$

where *BOOKNOS* is the set of all possible book numbers, *ISBNnos* is the set of all possible ISBN classification numbers, *titles* is the set of all possible book titles, and *AuthorNames* is the set of all possible author names.

Worked example 12.1 Write down a schema which describes the invariant properties of book details.

Solution The invariant property of book details is that each book has a title, an author, and an ISBN number. This can be described by the schema *InvBookDetails*

$$\begin{array}{|l}
\textit{InvBookDetails} \\
\hline
\textit{BookDetails} \\
\hline
\mathrm{dom}\ \textit{HasISBN} = \mathrm{dom}\ \textit{HasTitle} = \mathrm{dom}\ \textit{HasAuthor} \\
\hline
\end{array}$$

■

Changes to book details can be described by the schema $\Delta BookDetails$ where

$$\Delta BookDetails \mathrel{\widehat{=}} InvBookDetails' \wedge InvBookDetails$$

The schema $\Xi BookDetails$ describes events which leave the book details unaffected.

$$\begin{array}{|l}
\Xi \textit{BookDetails} \\
\hline
\Delta \textit{BookDetails} \\
\hline
\theta \textit{BookDetails}' = \theta \textit{BookDetails} \\
\hline
\end{array}$$

The schema *BookCategories* describes the fact that books can be missing, on the shelves of the library, or borrowed:

$$\begin{array}{|l}
\textit{BookCategories} \\
\hline
\textit{BorrowedBooks}, \textit{ShelfBooks}, \\
\textit{MissingBooks}, \textit{CatalogueBooks} : \mathbb{F}\ \textit{BOOKNOS} \\
\hline
\end{array}$$

where *CatalogueBooks* are the books in the library, *BorrowedBooks* are those that the system regards as being borrowed, *MissingBooks* are those books that the system regards as missing, and *ShelfBooks* are those books which the system regards as borrowable.

The next step is to specify the static or invariant properties of book categories. These properties must hold no matter what events occur in the library system. Although not explicitly stated in the specification, a book can only be borrowed, on the shelves, or missing. It cannot be a combination of these categories; for example, it cannot be borrowed *and* on the shelves.

Given these static properties, the schema *InvBookCategories* can be written which describes these properties.

$$
\begin{array}{|l}
\textit{InvBookCategories} \\
\hline
\textit{BookCategories} \\
\hline
\textit{BorrowedBooks} \cup \textit{ShelfBooks} \cup \textit{MissingBooks} = \textit{CatalogueBooks} \\
\textit{BorrowedBooks} \cap \textit{ShelfBooks} = \varnothing \\
\textit{ShelfBooks} \cap \textit{MissingBooks} = \varnothing \\
\hline
\end{array}
$$

Finally, the schemas $\Delta \textit{BookCategories}$ and $\Xi \textit{BookCategories}$ can be defined

$$\Delta \textit{BookCategories} \mathrel{\widehat{=}} \textit{InvBookCategories}' \wedge \textit{InvBookCategories}$$

$$
\begin{array}{|l}
\Xi \textit{BookCategories} \\
\hline
\Delta \textit{BookCategories} \\
\hline
\theta \textit{BookCategories}' = \theta \textit{BookCategories} \\
\hline
\end{array}
$$

Books can also be overdue. This can be described by the schema *overdues*.

$$
\begin{array}{|l}
\textit{overdues} \\
\hline
\textit{OverdueBooks} : \mathbb{F}\ \textit{BOOKNOS} \\
\hline
\end{array}
$$

Changes to overdue books and operations which do not affect overdue books can be defined by $\Delta \textit{overdues}$ and $\Xi \textit{overdues}$.

$$\Delta \textit{overdues} \mathrel{\widehat{=}} \textit{overdues}' \wedge \textit{overdues}$$

$$
\begin{array}{|l}
\Xi \textit{overdues} \\
\hline
\Delta \textit{overdues} \\
\hline
\textit{OverdueBooks}' = \textit{OverdueBooks} \\
\hline
\end{array}
$$

Given the schemas which describe book details and book categories, the part of the library system dealing with books can be specified. The schemas *books* and *InvBooks* can now be defined.

$$\begin{array}{|l}
\hline
\textit{books} \\
\hline
BookDetails \\
BookCategories \\
overdues \\
\hline
\end{array}$$

$$\begin{array}{|l}
\hline
\textit{InvBooks} \\
\hline
InvBookDetails \\
InvBookCategories \\
overdues \\
\hline
\mathrm{dom}\, HasTitle = CatalogueBooks \\
OverdueBooks \subseteq BorrowedBooks \\
\hline
\end{array}$$

The first predicate in *InvBooks* asserts that all the catalogue books have a title. Since

$$\mathrm{dom}\, HasISBN = \mathrm{dom}\, HasTitle = \mathrm{dom}\, HasAuthor$$

is asserted in the included schema *InvBookDetails*, this implies that every catalogued book has a title, an author name, and an ISBN number. $\Delta books$ and $\Xi books$ can now be specified as

$$\Delta books \mathrel{\widehat{=}} InvBooks' \wedge InvBooks$$

and

$$\begin{array}{|l}
\hline
\Xi Books \\
\hline
\Xi BookDetails \\
\Xi BookCategories \\
\Xi overdues \\
\hline
\end{array}$$

Users of the system can be described in a similar way. There are two types of users. These can be modelled by means of the schema *users*

$$\begin{array}{|l}
\hline
\textit{users} \\
\hline
staff, students, LibraryUsers : \mathbb{F}\ USERNOS \\
AddressOf : USERNOS \twoheadrightarrow addresses \\
NameOf : USERNOS \twoheadrightarrow UserNames \\
\hline
\end{array}$$

where *LibraryUsers* is the user community of the library, *UserNames* is the set of all possible user names, *addresses* is the set of all possible addresses, and *USERNOS* is the set of all possible user registration numbers.

AddressOf and *NameOf* are partial functions which relate a user to his address and name.

One invariant property of users is that, since there cannot be any other type of user, the collection of staff users and student users forms the whole user population. Another invariant property is that each user has both a name and an address.

Another possible property is that no user can be both a staff user and a student user. However, this is not explicitly stated in the statement of requirements. There is a possibility that a member of staff is studying for a higher degree *and* is registered with the University as a student. The statement of requirements does not deal with this. The originator of the statement of requirements and library staff will have to be asked how the system should deal with this. Assuming that a member of staff cannot be simultaneously a student user and a staff user, the schema describing invariant properties of users *InvUsers* can be defined.

Worked example 12.2 Write down the schema *InvUsers*.

Solution First, since a user can either be a member of staff or a student, the whole user population of the library will be the union of *staff* and *students*. Second, since a user cannot be both a member of staff and a student, the intersection of *staff* and *students* will be the empty set. Third, since each user has a name and an address, the domain of *AddressOf* will equal the domain of *NameOf*.

$$\begin{array}{|l}
\textit{InvUsers} \\
\hline
\textit{users} \\
\hline
\textit{staff} \cup \textit{students} = \textit{LibraryUsers} \\
\textit{staff} \cap \textit{students} = \varnothing \\
\operatorname{dom} \textit{AddressOf} = \operatorname{dom} \textit{NameOf} = \textit{LibraryUsers} \\
\hline
\end{array}$$

■

Finally, the two schemas $\Delta users$ and $\Xi users$ can be defined.

$$\Delta users \mathrel{\widehat{=}} InvUsers' \wedge InvUsers$$

$$\begin{array}{|l}\hline \Xi users \\ \Delta users \\ \hline \theta users' = \theta users \\ \hline\end{array}$$

To complete the description of the static properties it is necessary to describe the fact that users borrow books. The fact that a user is currently borrowing a book can be described by a relation *borrows* which is over $USERNOS \times BOOKNOS$. The fact that a book is borrowed on a certain day can be modelled by a partial function *DayBorrowed* which maps book numbers into days. These can be described by the schema *borrowings*. We shall assume that *days* are natural numbers measured from some base date, for example, from 1st Jan 1900.

$$\begin{array}{|l}\hline borrowings \\ borrows : USERNOS \nrightarrow \mathbb{F}\ BOOKNOS \\ DayBorrowed : BOOKNOS \nrightarrow days \\ \hline\end{array}$$

The schema *InvBorrowings* describes the fact that an invariant property of borrowings is that every book borrowed by a user is associated with a day on which it was borrowed. Also, that each book can only be borrowed by one user at a time.

$$\begin{array}{|l}\hline InvBorrowings \\ borrowings \\ \hline \mathrm{dom}\ DayBorrowed = \mathrm{ran}\ borrows \\ \forall\, u1, u2 : USERNOS \mid \{u1, u2\} \subseteq \mathrm{dom}\ borrows \bullet \\ \quad borrows\ us \cap borrows\ u2 \neq \varnothing \Rightarrow u1 = u2 \\ \hline\end{array}$$

The schema $\Delta borrowings$ which describes changes to borrowings can be described by

$$\Delta borrowings \mathrel{\widehat=} InvBorrowings' \wedge InvBorrowings$$

Any event which does not change the schema is described by $\Xi borrowings$ defined as

$$\begin{array}{|l}\hline \Xi borrowings \\ \Delta borrowings \\ \hline \theta borrowings' = \theta borrowings \\ \hline\end{array}$$

The part of the system dealing with reservations can be described by the schema *reservations*.

$$\begin{array}{|l}\hline \textit{reservations} \\ \textit{ReserveLists} : \textit{BOOKNOS} \nrightarrow \operatorname{seq} \textit{USERNOS} \\ \textit{DayReserve} : \mathbb{F}\ \textit{BOOKNOS} \\ \hline \end{array}$$

DayReserve will contain the book numbers of those books which have been reserved and returned to the library; these books will be regarded by the system as being on the shelves of the library. At the end of each day *DayReserve* will be used to notify users that reserved books await collection at the library. The invariant properties of the part of the system dealing with reservations is described by *InvReservations*.

$$\begin{array}{|l}\hline \textit{InvReservations} \\ \textit{reservations} \\ \hline \textit{DayReserve} \subseteq \{\ b : \textit{BOOKNOS} \mid \#(\textit{ReserveLists}\ b) > 0\ \} \\ \hline \end{array}$$

This states that the books that are regarded as being reserved and returned will be a subset of those books which have one or more users waiting for a book.

$\Delta reservations$ and $\Xi reservations$ can then be defined as

$$\Delta reservations \mathrel{\widehat{=}} InvReservations' \wedge InvReservations$$

and

$$\begin{array}{|l}\hline \Xi \textit{reservations} \\ \Delta \textit{reservations} \\ \hline \theta \textit{reservations}' = \theta \textit{reservations} \\ \hline \end{array}$$

The whole library system can then be partly described by the signature *sys*.

$$\begin{array}{|l}\hline \textit{sys} \\ \textit{books} \\ \textit{users} \\ \textit{borrowings} \\ \textit{reservations} \\ \hline \end{array}$$

The invariant properties which relate objects defined in *books*, *borrowings*, *overdues*, *reservations*, and *users* are:

- All reserved books are borrowed by other users or are missing.

- A student user may borrow no more than 10 books.
- A staff user may borrow no more than 20 books.
- All books taken out by users are regarded as borrowed.
- Each catalogued book is associated with a possibly non-empty reservation list.
- All borrowed books are borrowed by library users.
- All reserved books are reserved by library users.

There is a further property which may hold but which the statement of requirements does not address. This is the number of books that a user may reserve. Again the originator of the statement of requirements or his staff will need to be interrogated. We shall assume that the number of books reserved for any user will not be limited.

Worked example 12.3 Write down a schema *InvSys* which embodies both the properties specified above and the individual invariant properties of books, users, borrowings, and reservations.

Solution The schema is shown below

InvSys

$InvBooks\ InvUsers\ InvBorrowings\ InvReservations$

$\{\ b : BOOKNOS \mid (\#ReserveLists\ b) > 0\ \} \subseteq$
$BorrowedBooks \cup MissingBooks \cup DayReserve$

$\forall\, stu : students\ \bullet$
$\quad \#(borrows\ stu) \leq 10$

$\forall\, sta : staff\ \bullet$
$\quad \#(borrows\ sta) \leq 20$

$\mathrm{ran}\ borrows = BorrowedBooks$

$CatalogueBooks = \mathrm{dom}\ ReserveLists$

$\mathrm{dom}\ borrows \subseteq LibraryUsers$

$\forall\, book : CatalogueBooks\ \bullet$
$\quad \mathrm{ran}(ReserveLists\ book) \subseteq LibraryUsers$

$DayReserve \subseteq ShelfBooks \wedge OverdueBooks \subseteq BorrowedBoooks$

■

Given the schema *InvSys*, the schemas Δsys and Ξsys can be defined as

$$\Delta sys \mathrel{\widehat=} InvSys' \land InvSys$$

and

$$\begin{array}{|l}
\hline \Xi sys \\
\Xi books \\
\Xi users \\
\Xi reservations \\
\Xi borrowings \\
\hline
\end{array}$$

Finally, a schema *exceptions* can be defined. This describes the actions that are taken when an exception to normal processing occurs, for example, when an error occurs.

$$exceptions \mathrel{\widehat=} [\Xi sys;\ message! : ErrorTypes]$$

message! will have differing values when exceptions occur. It will give an indication of the type of exception that has occurred. Ξsys is used in the definition of *exceptions* because the system does not change when an error is discovered.

If a successful operation occurs, then *message*! will be given the value *ok*. This is defined by the schema *success*.

$$\begin{array}{|l}
\hline success \\
message! : ErrorTypes \\
\hline
message! = ok \\
\hline
\end{array}$$

12.3 The operations

A large bulk of the statement of requirements is concerned with the operation of the system. It deals with occurrences such as a user returning a book. This section describes how to model these by means of events and observations.

Paragraph 2.1. This paragraph describes the registration of a new user. There is a problem with this part of the statement of requirements: how do we know whether the user exists on the system given that we only know his or her name and address? There is a high probability that a user who has the same name and address as an existing user will be that user. However, in a university there are always a large number of shared houses and halls of residence where, say, two John Smiths may live. If this occurs then the library clerk has to discover whether the potential user already exists on the system. If so, then some form of admonishment is in order. If not, then a good strategy is to modify the user's address, for example by adding a room number, in the hope of making it unique and then re-register the details.

The schema which describes the fact that a typed name and address of a user may already exist in the library system is *AddandNamein*.

$$
\begin{array}{|l}
\textit{AddandNamein} \\
\hline
addr? : addresses \\
name? : UserNames \\
users \\
\hline
\exists\, usno : USERNOS \bullet \\
\qquad AddressOf\ usno = addr? \land NameOf\ usno = name? \\
\hline
\end{array}
$$

The schema *add* will partially describe the successful registration of a user into the library system. Using $\Xi books$, $\Xi reservations$, and $\Xi borrowings$ indicates that the event described by the schema does not affect books, borrowings, and reservations of the system.

$$
\begin{array}{|l}
\textit{add} \\
\hline
\Xi books \\
\Xi reservations \\
\Xi borrowings \\
\Delta users \\
success \\
user? : UserNames \\
UsNo? : USERNOS \\
UsAddr? : addresses \\
\hline
UsNo? \notin LibraryUsers \\
AddressOf' = AddressOf \cup \{UsNo? \mapsto UsAddr?\} \\
NameOf' = NameOf \cup \{UsNo? \mapsto user?\} \\
LibraryUsers' = LibraryUsers \cup \{UsNo?\} \\
\hline
\end{array}
$$

This can then be used in the schema *InsertStaff* which describes the insertion of staff details into the library system.

$$\begin{array}{|l}
\hline
\textit{InsertStaff} \\
\textit{add} \\
\hline
\textit{staff}' = \textit{staff} \cup \{ \textit{UsNo?} \} \\
\textit{students}' = \textit{students} \\
\hline
\end{array}$$

Similarly, the insertion of a student into the system can be described by the schema *InsertStudent*.

$$\begin{array}{|l}
\hline
\textit{InsertStudent} \\
\textit{add} \\
\hline
\textit{students}' = \textit{students} \cup \{ \textit{UsNo?} \} \\
\textit{staff}' = \textit{staff} \\
\hline
\end{array}$$

The error that occurs when duplicate names and addresses are discovered can be defined by the schema *ExistingUser*.

$$\begin{array}{|l}
\hline
\textit{ExistingUser} \\
\textit{AddandNameIn} \\
\textit{exceptions} \\
\hline
\textit{message!} = \textit{ExistingUserError} \\
\hline
\end{array}$$

A further error can occur if the library clerk has allocated a new number to the user which has already been allocated to another user. This can be described by the schema *ExistingNumber*.

$$\begin{array}{|l}
\hline
\textit{ExistingNumber} \\
\textit{UsNo?} : \textit{USERNOS} \\
\textit{exceptions} \\
\hline
\textit{UsNo?} \in: \textit{LibraryUsers} \\
\textit{message!} = \textit{ExistingNumberError} \\
\hline
\end{array}$$

The events that are associated with paragraph 2.1 of the specification can now be fully defined as

$$\begin{array}{l}
\textit{InsertStaff2.1} \mathrel{\widehat=} \\
\quad (\textit{InsertStaff} \land \neg\, \textit{AddandNameIn}) \lor \textit{ExistingUser} \lor \\
\quad \textit{ExistingNumber}
\end{array}$$

$$\begin{array}{l}
\textit{InsertStudent2.1} \mathrel{\widehat=} \\
\quad (\textit{InsertStudent} \land \neg\, \textit{AddandNameIn}) \lor \textit{ExistingUser} \\
\quad \lor\, \textit{ExistingNumber}
\end{array}$$

Paragraph 2.2. The removal of a student or staff user can be partly described by the schema *Remove*.

$$
\begin{array}{|l}
\textit{Remove} \\
\hline
\Xi borrowings \\
\Xi books \\
\Delta users \\
\Delta reservations \\
success \\
UsNo? : USERNOS \\
\hline
UsNo? \in LibraryUsers \\
\# borrows\ UsNo? = 0 \\
\forall book : \mathrm{dom}\ ReserveList \bullet \\
\quad ReserveList'\ book = squash(ReserveList\ book \rhd \{UsNo?\}) \\
LibraryUsers' = LibraryUsers \\
\hline
\end{array}
$$

For a user to be removed successfully he or she must be already registered with the library system and must no longer have any books out on loan. When a user is removed, all information about the books that he or she has reserved should be removed from the system.

This schema can be used in *RemoveStudent* and *RemoveStaff* which describes the act of removing a student and a member of staff from the system, respectively.

$$
\begin{array}{|l}
\textit{RemoveStudent} \\
\hline
remove \\
\hline
students' = students \setminus \{UsNo?\} \\
staff' = staff \\
\hline
\end{array}
$$

$$
\begin{array}{|l}
\textit{RemoveStaff} \\
\hline
remove \\
\hline
staff' = staff \setminus \{UsNo?\} \\
students' = students \\
\hline
\end{array}
$$

One error that can occur is when the user number is not recognized by the system.

$$
\begin{array}{|l}
\hline
\textit{InvalidUserNo} \\
UsNo? : USERNOS \\
exceptions \\
\hline
UsNo? \notin LibraryUsers \\
message! = NonExistentUserError \\
\hline
\end{array}
$$

Another error occurs when the user still has books out on loan.

$$
\begin{array}{|l}
\hline
\textit{BooksOnLoan} \\
UsNo? : USERNOS \\
exceptions \\
\hline
\#(borrows\ UsNo?) > 0 \\
message! = BooksStillBorrowedError \\
\hline
\end{array}
$$

Paragraph 2.2 can now be fully defined as

$$RemoveStudent2.2 \mathrel{\widehat=} RemoveStudent \lor InvalidUsNo \lor BooksOnLoan$$

$$RemoveStaff2.2 \mathrel{\widehat=} RemoveStaff \lor InvalidUsNo \lor BooksOnLoan$$

Paragraph 2.3. This paragraph is concerned with the change of address and name of user.

Worked example 12.4 Specify the effect of paragraph 2.3.

Solution Both the operations will leave the books in the library and the objects described in *borrowings* and *reservations* unchanged. A new address or new name is supplied together with the registration number of a user. This registration number must match one stored in the library system. The schema *change* partly describes the effect of a successful change of address or change of name.

$$
\begin{array}{|l}
\hline
\textit{change} \\
\Xi books \\
\Xi borrowings \\
\Xi reservations \\
\Delta users \\
success \\
UsNo? : USERNOS \\
\hline
UsNo? \in LibraryUsers \\
\hline
\end{array}
$$

The schemas *ChangeAddress* and *ChangeName* specify the effect of a successful name change and address change.

$$\begin{array}{|l}
\hline
\textit{ChangeName} \\
\textit{change} \\
\textit{NewName?} : \textit{UserNames} \\
\hline
\textit{NameOf}' = \textit{NameOf} \oplus \{(\textit{UsNo?}, \textit{NewName?})\} \\
\hline
\end{array}$$

$$\begin{array}{|l}
\hline
\textit{ChangeAddress} \\
\textit{change} \\
\textit{NewAddress?} : \textit{UserAddresses} \\
\hline
\textit{AddressOf}' = \textit{AddressOf} \oplus \{(\textit{UsNo?}, \textit{NewAddress?})\} \\
\hline
\end{array}$$

The only error that can occur with these events is when an invalid user registration number is typed by the library assistant. This has already been described in the schema *InvalidUsNo*? which was used to describe the event associated with paragraph 2.2. Paragraph 2.3 can hence be specified as

$$\textit{ChangeName}2.3 \mathrel{\widehat{=}} \textit{ChangeName} \lor \textit{InvalidUsNo}$$

$$\textit{ChangeAddress}2.3 \mathrel{\widehat{=}} \textit{ChangeAddress} \lor \textit{InvalidUsNo}$$

The single predicates cover both the change of name and address.
■

Paragraph 2.4. This paragraph describes the insertion of book details into the library system. The library assistant provides: an ISBN number, a book title, and a registration number. The schema *InsertBook* shown on the next page describes this process.

The only error that can occur when the operation is carried out is a book registration number being provided that is already in use. This can be described by the schema *InvalidBookNo* which is again shown on the next page. Paragraph 2.4 can then be specified as

$$\textit{InsertBook}2.4 \mathrel{\widehat{=}} \textit{InsertBook} \lor \textit{InvalidBookNo}$$

InsertBook

$\Xi users\ \Xi borrowings\ \Xi reservations\ \Xi overdues$
$\Delta BookCategories\ \Delta BookDetails$
$successNewISBN? : ISBNNos$
$NewTitle : titles$
$NewBookNo? : BOOKNOS$
$NewAuthor? : AuthorNames$

$NewBookNo? \notin CatalogueBooks$

$HasISBN' = HasISBN \cup \{NewBookNo? \mapsto NewISBN?\}$

$HasTitle' = HasTitle \cup \{NewBookNo? \mapsto NewAuthor?\}$

$MissingBooks' = MissingBooks$

$BorrowedBooks' = BorrowedBooks$

$ShelfBooks = ShelfBooks \cup \{NewBookNo?\}$

$CatalogueBooks' = CatalogueBooks \cup \{NewBookNo?\}$

InvalidBookNo

$NewBookNo? : BOOKNOS$
$exceptions$

$NewBookNo? \in CatalogueBooks$

$message! = InvalidBookRegistrationNumberError$

Paragraph 2.5. The operation of generating unused registration numbers does not change the library system. It forms a sequence of book numbers which are in ascending order. The range of the sequence is a subset of those natural numbers less than a million which are not already catalogue numbers. The schema which describes the successful event that corresponds to this operation is *ProvideBookNos* In this schema *NoOfBookNumbers*? is the number of book registration numbers required. The first predicate states that a maximum of 1000 book numbers are to be provided by the library assistant. The second predicate states that the number of unused numbers must be greater than or equal to the number requested. The third predicate establishes the fact that they will be unused. The fourth predicate specifies that the sequence *UnusedBookNumbers*! will be in ascending order. Finally, the fifth predicate describes the fact that the number of book registration numbers requested will be provided.

$$
\begin{array}{|l}
\hline
\textit{ProvideBookNos} \\
\Xi sys \\
success \\
NoOfBookNumbers? : \mathbb{N} \\
UnusedBookNumbers! : \mathrm{seq}\ \mathbb{N} \\
\hline
NoOfBookNumbers? \leq 1000 \\
\#\{\ a : \mathbb{N} \mid a < 1\,000\,000 \land a \notin CatalogueBooks\ \} \\
\quad \geq NoOfBookNumbers? \\
\mathrm{ran}\ UnusedBookNumbers! \subseteq \\
\quad \{\ a : \mathbb{N} \mid a < 1\,000\,000 \land a \notin CatalogueBooks\ \} \\
\forall i, j : \mathrm{dom}\ UnusedBookNos? \bullet \\
\quad i < j \Rightarrow UnusedBookNos!\ i < UnusedBookNos!\ j \\
\#UnusedBookNumbers! = NoOfBookNumbers? \\
\hline
\end{array}
$$

The first error that can occur is when the library assistant types in a request to provide more than 1000 unused registration numbers. This can be described by the schema *InvalidNoOfBooks*.

$$
\begin{array}{|l}
\hline
\textit{InvalidNoOfBooks} \\
NoOfBookNumbers? : \mathbb{N} \\
exceptions \\
\hline
NoOfBookNumbers? > 1000 \\
message! = TooManyBookNosError \\
\hline
\end{array}
$$

The second error that can occur is when not enough unused book numbers remain to satisfy the query.

$$
\begin{array}{|l}
\hline
\textit{NotEnoughBookNos} \\
NoOfBookNos? : \mathbb{N} \\
exceptions \\
\hline
\#\{\ a : \mathbb{N} \mid a < 1\,000\,000 \land a \notin CatalogueBooks\ \} \\
\quad < NoOfBookNumbers? \\
message! = TooFewBookNosAvailableError \\
\hline
\end{array}
$$

The effect of paragraph 2.5 can now be specified as

$$
\begin{array}{l}
ProvideBookNos2.5\ \widehat{=} \\
\quad ProvideBookNos \lor InvalidNoOfBooks \lor NotEnoughBookNos
\end{array}
$$

Paragraph 2.6. This is similar to paragraph 2.5. Consequently the schema will be similar.

Worked example 12.5 Specify the event described by paragraph 2.6 of the statement of requirements.

Solution The event that occurs when unused user registration numbers are produced is described by the schema *ProvideUserNos*.

$$
\begin{array}{|l}
\textit{ProvideUserNos} \\
\hline
\Xi sys \\
success \\
NoOfUserNumbers? : \mathbb{N} \\
UnusedUserNumbers! : \mathrm{seq}\ \mathbb{N} \\
\hline
NoOfUserNumbers? \leq 1000 \\
\#\{\ a : \mathbb{N} \mid a < 10\,000 \land a \notin LibraryUsers\ \} \geq \\
\quad NoOfUserNumbers? \\
\mathrm{ran}\ UnusedUserNumbers! \subseteq \\
\quad \{\ a : \mathbb{N} \mid a < 10\,000 \land a \notin LibraryUsers\ \} \\
\forall i, j : \mathrm{dom}\ UnusedUserNumbers! \bullet \\
\quad i < j \Rightarrow UnusedUserNumbers!\ i < UnusedUserNumbers!\ j \\
\#\, UnusedUserNumbers! = NoOfUserNumbers? \\
\hline
\end{array}
$$

The first error associated with this event occurs when the library assistant requests more than 1000 unused user numbers.

$$
\begin{array}{|l}
\textit{InvalidNoOfUsers} \\
\hline
NoOfUserNumbers? : \mathbb{N} \\
exceptions \\
\hline
NoOfUserNumbers? > 1000 \\
message! = TooManyUserNos \\
\hline
\end{array}
$$

The second error occurs when not enough user numbers are available to satisfy the request.

$$
\begin{array}{|l}
\hline
\textit{NotEnoughUserNos} \\
\textit{NoOfUserNumbers?} : \mathbb{N} \\
\textit{exceptions} \\
\hline
\#\{\ a : \mathbb{N} \mid a < 10\,000 \land a \notin \textit{LibraryUsers}\ \} < \\
\quad \textit{NoOfUserNumbers?} \\
\textit{message!} = \textit{TooFewUserNumbersAvailableError} \\
\hline
\end{array}
$$

Paragraph 2.6 can be fully specified as

$$
\begin{array}{l}
\textit{ProvideUserNos2.6} \mathrel{\widehat{=}} \\
\quad \textit{ProvideUserNos} \lor \textit{InvalidNoOfUsers} \lor \textit{NotEnoughUserNos}
\end{array}
$$

■

Paragraph 2.7. This part of the statement of requirements implies that an ordering of entries for the microfiche hard copy is required. This means that a sequence has to be used to model the individual entries in the catalogue. These elements can be described by the schema *MicroficheElements*.

$$
\begin{array}{|l}
\hline
\textit{MicroficheElements} \\
\textit{isbn} : \textit{ISBNNos} \\
\textit{title} : \textit{titles} \\
\textit{auth} : \textit{AuthorNames} \\
\textit{book} : \textit{BOOKNOS} \\
\textit{books} \\
\hline
\textit{book} \in \textit{CatalogueBooks} \\
\textit{isbn} = \textit{HasISBN book} \\
\textit{title} = \textit{HasTitle book} \\
\textit{auth} = \textit{HasAuthor book} \\
\hline
\end{array}
$$

The set from which elements of the microfiche sequence are taken is then defined by

$$
\begin{array}{l}
\textit{MicroficheEntries} == \\
\quad \{\ \textit{MicroficheElements} \bullet (\textit{book}, \textit{isbn}, \textit{title}, \textit{auth})\ \}
\end{array}
$$

The microfiche catalogue can be modelled by a sequence of elements which can be described by *MicroficheEntries*. The effect of constructing the hard copy for the catalogue can be specified by the schema *FormMicrofiche*.

$$
\begin{array}{|l}
\textit{FormMicrofiche} \\
\hline
\Xi sys \\
success \\
MicroficheCopy! : \mathrm{seq}\ MicroficheEntries \\
\hline
\mathrm{ran}\ MicroficheCopy! = MicroficheEntries \\
\\
\forall j, k : \mathrm{dom}\ MicroficheCopy! \bullet \\
j < k \Rightarrow \lambda isbn\ MicroficheCopy!\ j < \lambda isbn\ MicroficheCopy!\ k \\
\hline
\end{array}
$$

The predicate on the first line states that all the items in the microfiche hard copy will represent books which are in the library catalogue. The predicate written on the second and third lines describes the ordering outlined in paragraph 2.7. There will be no errors associated with this event. In the predicate the projection function *isbn* is used to extract elements from the tuples which make up the range of the sequence *MicroficheCopy*!.

Paragraph 2.12. The event described by paragraph 2.12 produces a list of remainders for all those library books which are overdue. Apart from the overdue books, it does not affect the library system. The statement of requirements does not specify any ordering of the reminder list; therefore, a set is chosen to represent it. The schema *FormsOverdueList* describes the successful execution of this event, *day*? is the day on which the event occurs. No errors are associated with this event.

$$
\begin{array}{|l}
\textit{FormsOverdueList} \\
\hline
\Xi users\ \Xi reservations\ \Xi borrowings \\
\Xi BookDetails\ \Xi BookCategories \\
\Delta overdues \\
success \\
BooksFoundOverdue : \mathbb{P}\ BOOKNOS \\
day? : days \\
OverdueRems! : \mathbb{P}\,(UserNames \times addresses \times titles \times BOOKNOS) \\
\hline
BooksFoundOverdue = \{\ no : BOOKNOS \mid \\
\quad no \in BorrowedBooks \land day? > DayBorrowed\ no + 30 \\
\quad \land\ no \notin OverdueBooks\ \} \\
\\
OverdueRems! = \{\ u : UserNames;\ a : addresses;\ t : titles; \\
\quad b : BOOKNOS \mid b \in BooksFoundOverdue \land \\
\quad \exists\, un : \mathrm{dom}\ borrows \bullet b \in borrows\ un \land u = NameOf\ un \land \\
\quad \exists\, un : \mathrm{dom}\ borrows \bullet b \in borrows\ un \land a = AddressOf\ un \land \\
\quad t = HasTitle\ b\ \} \\
\\
OverdueBooks' = OverdueBooks \cup BooksFoundOverdue \\
\hline
\end{array}
$$

Paragraph 2.21. The successful borrowing of a book can be described as the schema *BookBorrowed*.

$$
\begin{array}{l}
\textit{BookBorrowed} \\
\hline
\Xi users\ \Xi reservations\ \Xi BookDetails\ \Xi overdues \\
\Delta BookCategories\ \Delta Borrowings \\
success \\
day? : days,\ UsNo? : USERNOS,\ BookNo? : BOOKNOS \\
\hline
UsNo? \in LibraryUsers \\
BookNo? \in ShelfBooks \\
\#(ReserveLists\ BookNo?) = 0 \\
UsNo? \in staff \Rightarrow \#(borrows\ UsNo?) \leq 20 \\
UsNo? \in students \Rightarrow \#(borrows\ UsNo?) \leq 10 \\
ShelfBooks' = ShelfBook \setminus \{BookNo?\} \\
BorrowedBooks' = BorrowedBooks \cup \{BookNo?\} \\
MissingBooks' = MissingBooks \\
CatalogueBooks' = CatalogueBooks \\
DayBorrowed' = DayBorrowed \cup \{BookNo? \mapsto day?\} \\
borrows' = borrows \cup \{UsNo? \mapsto BookNo?\} \\
\hline
\end{array}
$$

For this event to occur successfully the user must be a registered library user, the book must be on the shelves, no user must have reserved the book to be borrowed, and the user must have borrowed less than his or her limit.

There are a number of errors associated with the borrowing of a book. First, a user may not be recognized by the system. This has already been described by the schema *InvalidUsNo* which was used to describe paragraph 2.2 of the statement of requirements. Second, a borrower could have borrowed the limit of books.

$$
\begin{array}{l}
\textit{TooManyBooks} \\
\hline
exceptions \\
UsNo? : USERNOS \\
\hline
(UsNo? \in staff \land \#(borrows\ UsNo?) = 20 \Rightarrow \\
\qquad message! = StaffOverLimitError) \lor \\
(UsNo? \in students \land \#(borrows\ UsNo?) = 10 \Rightarrow \\
\qquad message! = StudentOverLimitError \\
\hline
\end{array}
$$

Third, a book could already have been borrowed:

$$\begin{array}{|l} \hline \textit{AlreadyBorrowed} \\ \textit{exceptions} \\ BookNo? : BOOKNOS \\ \hline BookNo? \in BorrowedBooks \\ message! = BookBorrowedError \\ \hline \end{array}$$

Fourth, a book could be missing:

$$\begin{array}{|l} \hline \textit{MissingBook} \\ \textit{exceptions} \\ BookNo? : BOOKNOS \\ \hline BookNo? \in MissingBooks \\ message! = BookMissingError \\ \hline \end{array}$$

Fifth, a book could be reserved. In this case a message will be displayed, the book put to one side, and the book marked as awaiting collection. This can be described by the schema *BookReserved*.

$$\begin{array}{|l} \hline \textit{BookReserved} \\ \Xi books \\ \Xi users \\ \Xi borrowings \\ \Delta reservations \\ message! : ErrorTypes \\ BookNo? : BOOKNOS \\ \hline \#(ReserveLists\ BookNo?) > 0 \\ DayReserve' = DayReserve \cup \{BookNo?\} \\ ShelfBooks' = ShelfBooks \cup \{BookNo?\} \\ message! = BookAlreadyReserved \\ \hline \end{array}$$

Paragraph 2.21 can now be fully specified as

$$BookBorrowed2.21 \mathrel{\widehat{=}} BookBorrowed \lor InvalidUsNo \lor TooManyBooks \lor AlreadyBorrowed \lor MissingBook \lor BookReserved$$

Paragraph 2.22. The successful return of a book can be described by the schema *BookReturned*.

$$
\begin{array}{|l}
\hline
\textit{BookReturned} \\
\Xi users \\
\Xi BookDetails \\
\Xi reservations \\
\Xi overdues \\
\Delta BookCategories \\
\Delta borrowings \\
success \\
UsNo? : USERNOS \\
BookNo? : BOOKNOS \\
\hline
\#(ReserveLists\ BookNo?) = 0 \\
UsNo? \in LibraryUsers \\
BookNo? \in BorrowedBooks \\
BookNo? \notin OverdueBooks \\
(UsNo?, BookNo?) \in borrows \\
BorrowedBooks' = BorrowedBooks \setminus \{BookNo?\} \\
ShelfBooks' = ShelfBooks \cup \{BookNo?\} \\
MissingBooks' = MissingBooks \\
CatalogueBooks' = CatalogueBooks \\
DayBorrowed' = \{BookNo?\} \lhd DayBorrowed \\
Borrows' = Borrows \rhd \{BookNo?\} \\
\hline
\end{array}
$$

The first five predicates establish the conditions that are to be true if a successful book return occurs. The first predicate states that the book must not be reserved; the second predicate states that the user must be a recognized library user; the third predicate states that the book must have been borrowed; the fourth predicate states that the book must not be overdue; finally, the fifth predicate states that the user returning the book must be the user who borrowed the book.

The sixth and seventh predicates establish the fact that the book is no longer borrowed and is on the shelves. The eight and ninth predicates establish the fact that a successful borrowing does not affect the list of missing books and catalogued books. The tenth and eleventh predicates remove any trace of the fact that a book has been borrowed.

The first exception that is associated with the event occurs when the book returned is regarded by the system as being on the shelves of the library. This might occur when a library assistant forgot to notify the system when the book was borrowed.

BookOnShelves

$exceptions$
$BookNo? : BOOKNOS$

$BookNo? \in ShelfBooks$

$message! = NotBorrowedBookError$

Second, the book may be regarded by the system as missing. This is described by the schema *MissingBook* already defined in connection with paragraph 2.21. Third, the user registration number may not be recognized by the system. This is described by the schema *InvalidUsNo* described in connection with paragraph 2.2, and *BookReserved* defined in connection with paragraph 2.21. Fourth, the user who returns a book may not be the user who borrowed the book. A message should be displayed for this and the library assistant should take appropriate action, for example, finding out who actually borrowed the book and keying in the correct user registration number. This exception can be described by the schema *WrongBorrower*.

WrongBorrower

$UsNo? : USERNOS$
$BookNo? : BOOKNOS$
$exceptions$

$(UsNo? \mapsto BookNo?) \notin borrows$

$message! = WrongBorrowerError$

Another important exception that can occur is a book being returned which is overdue. This can be specified by the schema *OverdueBook*.

OverDueBook

$\Xi users\ \Xi BookDetails\ \Xi borrowings$
$\Xi reservations\ \Xi BookCategories$
$\Delta overdues$
$BookNo? : BOOKNOS$

$BookNo? \in OverdueBooks$

$OverdueBooks' = OverdueBooks \setminus \{BookNo?\}$

$message! = BookOverdueError$

The library assistant will normally fine the borrower and then inform the system that the book has been returned. The event of returning a book can now be fully specified as

$$\begin{aligned} &BookReturned2.22 \mathrel{\widehat=} BookReturned \lor BookOnShelves \lor \\ &\quad BookMissing \lor InvalidUsNo \lor BookReserved \\ &\quad \lor WrongBorrower \lor OverdueBook \end{aligned}$$

Paragraph 2.30. This paragraph is relatively easy to specify. The successful operation of notifying the system that a missing book has been found is described by the schema *NotifyFoundBook*.

```
┌─ NotifyFoundBook ──────────────────────────
│ Ξusers
│ ΞBorrowings
│ Ξreservations
│ ΞBookDetails
│ ΞBorrowings
│ ΔBookCategories
│ BookNo? : BOOKNOS
├──────────
│ BookNo? ∈ MissingBooks
│ #(ReserveLists BookNo?) = 0
│ ShelfBooks' = ShelfBooks ∪ {BookNo?}
│ MissingBooks' = MissingBooks \ {BookNo?}
│ CatalogueBooks' = CatalogueBooks
│ BorrowedBooks' = BorrowedBooks
└────────────────────────────────────────────
```

Worked example 12.6 What errors can occur when a library assistant notifies the system that a missing book has been found?

Solution Two errors can occur: the book could already be marked as on the library shelves or it could be marked as borrowed.
■

The errors described in the previous worked example can be described by the schemas *ShelfBookFound* and *BorrowedBookFound*.

$$
\begin{array}{|l}
\hline
\textit{ShelfBookFound} \\
\textit{exceptions} \\
\textit{BookNo?} : \textit{BOOKNOS} \\
\hline
\textit{BookNo?} \in \textit{ShelfBooks} \\
\textit{message!} = \textit{BookAlreadyOnShelves} \\
\hline
\end{array}
$$

$$
\begin{array}{|l}
\hline
\textit{BorrowedBookFound} \\
\textit{exceptions} \\
\textit{BookNo?} : \textit{BOOKNOS} \\
\hline
\textit{BookNo?} \in \textit{BorrowedBooks} \\
\textit{message!} = \textit{BookAlreadyBorrowed} \\
\hline
\end{array}
$$

The missing book could be reserved for a user. In this case the library assistant should keep the book and the system notified. This can be achieved by the schema *BookReserved* described previously. The event of notifying a missing book can now be fully specified as

$$
\begin{array}{l}
\textit{NotifyFoundBook}2.30 \mathrel{\widehat{=}} \textit{NotifyFoundBook} \vee \textit{ShelfBookFound} \\
\qquad \vee\ \textit{BorrowedBookFound} \vee \textit{BookReserved}
\end{array}
$$

Paragraph 2.35. The successful completion of the event which describes the sending out of notification letters at the end of the working day is described by the schema *NotificationReturn*.

$$
\begin{array}{|l}
\hline
\textit{NotificationReturn} \\
\Xi \textit{BookDetails} \\
\Xi \textit{users} \\
\Xi \textit{Borrowings} \\
\Xi \textit{BookCategories} \\
\Delta \textit{reservations} \\
\textit{notf!} : \mathbb{P}\,(\textit{UserNames} \times \textit{addresses} \times \textit{titles} \times \textit{BOOKNOS}) \\
\hline
\textit{notf!} = \\
\qquad \{\ u : \textit{UserNames};\ a : \textit{addresses};\ t : \textit{titles};\ b : \textit{BOOKNOS} \mid \\
\qquad b \in \textit{DayReserve} \wedge t = \textit{HasTitle}\ b\ \wedge \\
\qquad \exists\, \textit{un} : \mathrm{dom}\,\textit{borrows} \bullet b \in \textit{borrows}\ \textit{un} \wedge a = \textit{AddressOf}\ \textit{un}\ \wedge \\
\qquad \exists\, \textit{un} : \mathrm{dom}\,\textit{borrows} \bullet b \in \textit{borrows}\ \textit{un} \wedge u = \textit{NameOf}\ \textit{un} \} \\
\textit{DayReserve}' = \varnothing \\
\forall\, \textit{book} : \textit{DayReserve} \bullet \\
\qquad \textit{ReserveLists}'\ \textit{book} = \textit{tail}\ \textit{ReserveLists}\ \textit{book} \\
\hline
\end{array}
$$

The first predicate describes the set of all notifications to users. The second predicate describes the fact that at the end of the day the list of reserved books is cleared. The third predicate states that after the operation has been successfully completed all the queues for the books in the *DayReserve* set will have their first element removed.

13

Z and design

Aims

- To describe the difference between functional specification and design.
- To show one use of mathematical proof within formal software development.

13.1 Introduction

The previous chapters have described the process whereby a system is specified in terms of mathematics, using a language for specifying stored data, data invariants, pre-conditions, and post-conditions. A point which was made at the beginning of the book was that discrete mathematics is an excellent medium for specifying a system, because it just describes the bare bones of that system. A feature of the specifications that have been presented in the last two chapters is that they do not contain any design directives: the behaviour of the system is specified in terms of operations on mathematical structures, many of which are not implemented in programming languages.

For system specification this is ideal: you do not want design issues clouding the process of exploring the detailed functionality of a system. However, a time soon arrives in a software project when a design has to be considered, although it must be said that this should normally occur after system specification has been completed.

What this chapter aims to do is to look at what design involves, show how the mathematical language Z presented in the previous chapters can be used to specify the design of a system and, finally, describe some techniques for achieving good designs. The final part of the chapter will outline how a design can be validated against a top-level specification.

13.2 The nature of design

System designers are faced with one of the most intellectually difficult problems in computing. They are given a specification of a system which contains a description of the functions of that system, and then have to design it in terms of operations which not only satisfy the functional part of the specification, but also respect any response time and memory constraints contained in that document. Moreover, the designer has to produce a system that is not too complex in programming terms.

This last point is often forgotten: a software developer is often under severe budget and time constraints and a system which contains complex program code will soak up resources for testing and programming. This is particularly important for systems which are to undergo extensive maintenance. As much as 75% of software budgets are now spent on this activity, and a delivered system which contains complex code rapidly becomes a major drain on the resources of a software developer.

There is another added dimension to the design process. Let us assume that the designer of a system is given the specification of the system in terms of operations on some stored data. Some of these operations might be executed frequently, while others may only be executed a small number of times, for example, when the system is initialized. The designer has to chose a design which takes advantage of this information so that the design may result in infrequent operations being executed very slowly, with frequent operations being the fastest.

In order to illustrate the points that have been made, consider a small problem: that of the design of a set of integers. A number of operations are to be designed. These operations will insert an integer in the set, delete an integer from the set, count the number of items in the set, and check whether an item is in the set.

There are a number of ways in which this set can be implemented. For the purposes of the discussion of design let us assume that the set is to be implemented as an array, and consider some of the design variations that can be applied to this basic data type.

First, let us design the data with a linear search mechanism in mind. For simplicity's sake, assume that the array that is to hold the set members will contain no more than a hundred items. Linear search consists of a step-by-step search of the elements of the array, normally starting at the first element, and terminating either when the element to be searched for is found, or when the final element of the array is encountered. This is shown in Figure 13.1. Linear search has a number of advantages. First, it requires the minimum amount of storage: each item is stored in exactly one location of the array. Second, many of the operations are easy to program: if a variable n is used to hold the number of current items in the array,

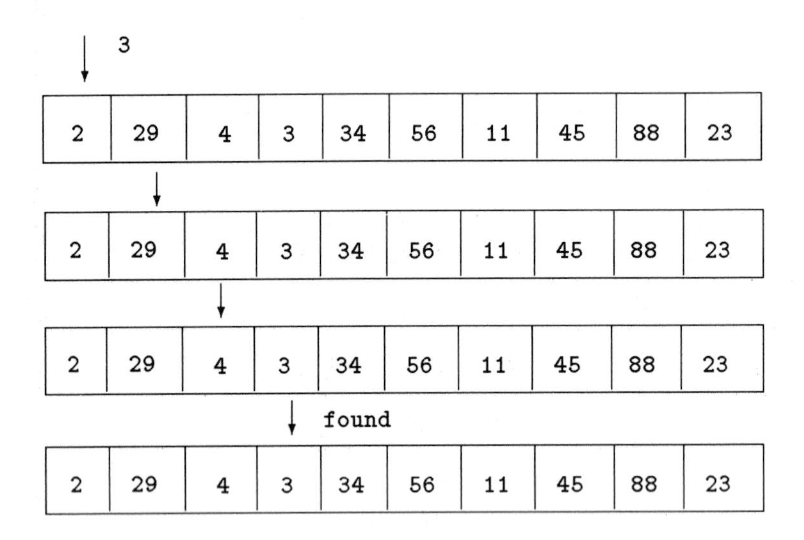

Figure 13.1 An example of linear search

then all insertion consists of is the placement of the item to be inserted at the $(n + 1)$th position in the array. Finding the number of items in the array is even easier: all it entails is reading the variable n. Deletion is more complex: the item to be deleted has to be found; then all the items after it have to be moved upward in order to remove all trace of the item.

The problem with linear search is that response time is poor. The average search time for a table organized for linear search is of the order of n, where n is the number of items in the table. This is certainly not the optimum response time that can be achieved with a set designed as an array.

So, a set designed as an array, organized for linear search, gives excellent memory utilization, low complexity of programming, and a high response time. Normally, such a design is selected when the array contains a small number of items (usually less than ten).

One way of improving a linear search is to arrange that when an item is retrieved it is placed at the front of the array, and all the other items in the array are shifted down by one, towards the item's old position. This is shown in Figure 13.2. If a small number of items are retrieved frequently then this tactic can speed up a linear search considerably.

Another way of organizing an array which implements a set is to arrange that the array is sorted, and apply a binary search. Here, the middle item of the array is examined. If the integer to be searched for lies be-

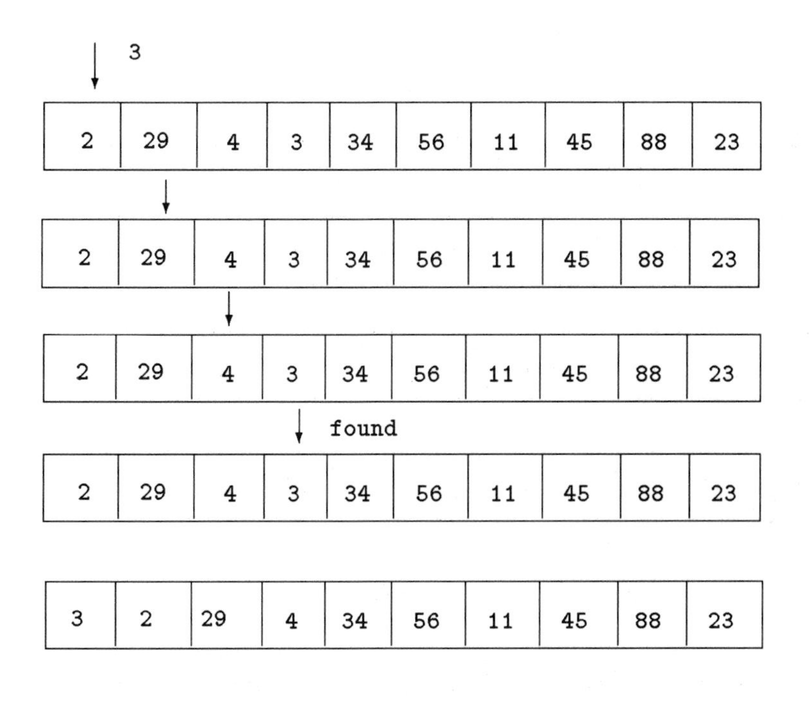

Figure 13.2 Augmented linear search

fore this middle element, then we know that it is in the upper part of the array, otherwise it is in the lower half. The middle element of whatever table has been identified is then examined, and the same decision made again, until the item is found as a middle element. This is shown in Figure 13.3. Binary search is fast—very fast—it runs in a time proportional to $\ln n$. However, there is a cost for achieving this speed-up. The cost is not in terms of memory: the array still holds no more than the integers to be searched. However, the programming complexity has increased, and some of the operations will be slower compared with a linear search. The programming complexity has increased because, since the array is ordered, it means, for example, that when an insertion is carried out the inserted integer is not placed at the end of the array, as with linear search, but is placed in the position that respects the ordering of the array. Insertion is also a little more expensive in terms of run-time, as compared with, say, a linear organization, because a loop is employed to find the insertion point. With the array organized for linear search the insertion occurs at the end of the array and would just require one assignment statement.

Another form of search is hashed search. Here, the position of an item

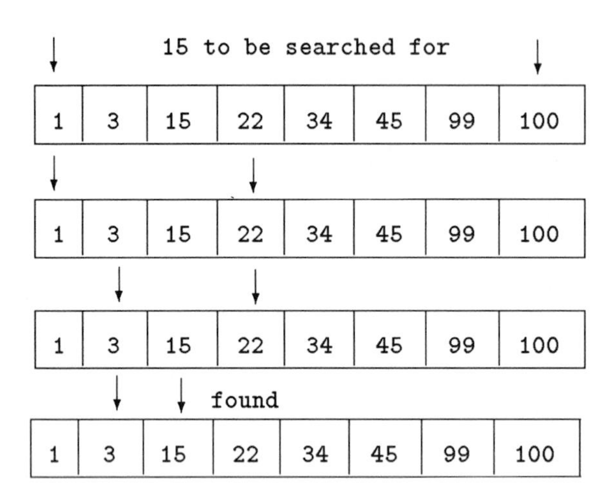

Figure 13.3 The process of binary search

in an array is determined by a function known as a *hash function*. This function, given either the item to be inserted or the key of the item, calculates an integer which lies within the bounds of the array which is to be used to store the item. So, for example, if integers were to be inserted into an array which had thirty elements starting at element 1, then a suitable hash function would take the remainder when the integer is divided by 30, and then add 1.

When an integer is to be inserted the hash function would be applied to the integer. The location generated would then be examined: if it was empty the location would be filled with the integer and the insertion process terminated. However, if the location already contained an integer—this might happen since a hash function will not generate a unique location for its argument—then the next empty location is searched for. There are a number of ways of determining the next empty location. One simple way is to carry out a linear search from the failed insertion point and find the first empty location. If the linear search reached the end of the array without an empty location being found, then the search re-commences at the start of the table. The insertion process is shown in Figure 13.4 for three integers being inserted into an array which has ten entries.

Let us assume that we want to retrieve an integer from our hashed array of integers. How would this be done? First, the hash function is applied to the integer to give an initial position in the array. The items in the array are then examined from that position in linear sequence until either one

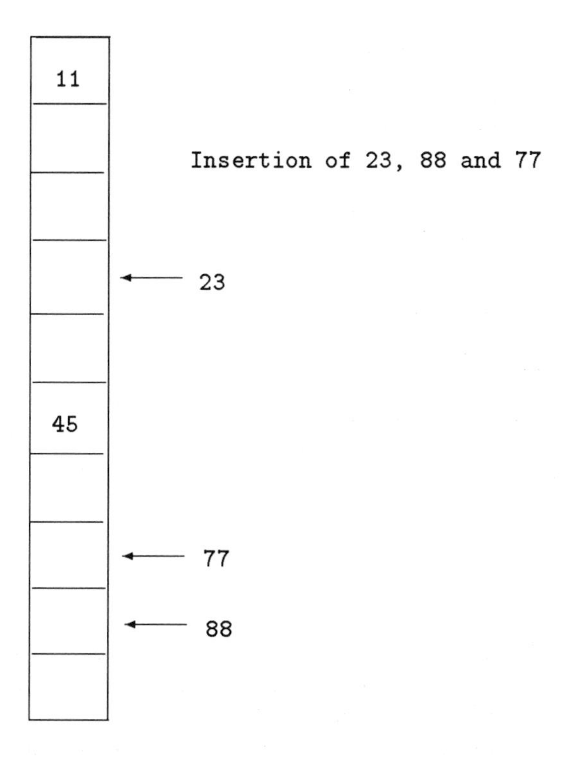

Figure 13.4 Insertion in a hashed table

of two conditions occurs: first, an empty location is encountered, in which case the integer is not in the array or, second, the integer is found, in which case it is in the table.

Hashing is a very fast way of organizing an array. A judicious choice of hashing function often means that for most of the time the initial location generated will contain the item that is to be searched for. Its disadvantage is that for a rapid response time quite a number of locations in an array need to be empty—something like 40%.

This, then, is an outline of hashed search. It is very fast, is not optimal in terms of memory, and is slightly difficult to program in that the search from the bottom of the table to the top can give rise to errors which can be difficult to detect.

The search techniques which have been described exemplify all the trade-offs inherent in designing a system: memory usage, response time and programming complexity. There are a number of design heuristics which can be used in design. The next section details them.

13.3 Some design heuristics

There are a number of standard strategies which can be used to achieve a good design in terms of memory, response time, or programming complexity, and the aim of this section is to describe them.

13.3.1 Pre-computing results

Often there is a need for an operation which calculates some result from stored data. As a simple example consider the system which receives receipts from salesmen about the sales that have been made by them. There is often a requirement for such systems to provide a facility whereby the average amount of sales is to be displayed on some output device such as a VDU. If the sales are held in an array, then all this would require is the scanning of the array item-by-item, with a cumulative total being formed. This is then divided by the number of items in the array. For an array of n entries the time taken will be proportional to n.

If the average command is used very frequently, then an alternative strategy would be to supplement the sales table with a variable which held the current average. When the average is required, then all that is needed is to read this variable. In order to achieve this, the variable has to be updated whenever a new sale is entered or deleted. However, if insertion and deletion occur relatively infrequently the response time of the average command will be considerably improved at a small cost: the cost of an extra memory location which holds the current average, and an extra calculation in the code fragments which implement the insertion and deletion operations on the sales data base.

Another example of this strategy in action again occurs in the implementation of sets. Assume that we have implemented a set as an array which stores names for a commercial data processing application such as a stock control system. Also, assume that there are two common operations in the system as well as the standard insertion and deletion operations. The first operation returns with the number of items in the array, and the second checks if the array is full. If the set is implemented as a hashed array, then a simple implementation of these operations would require a scan of the array, item-by-item, counting up the number of names in the array.

However, if an extra variable is used which holds the current number of items in the array, then all that is required for the operation which returns the number of items in the array, is to return the value of the variable; and all that is required for the operation which checks whether the array is empty, is to return a boolean formed by comparing the variable containing the number of items with zero.

This speed-up again has a cost: an increase of one memory location to hold the number of items in the table, and a small amount of extra processing in any operations which insert elements in the array or delete elements from the array.

13.3.2 Storing frequently used data

This is another popular technique for reducing response time, often at little cost. Many applications which retrieve data often access a limited subset of that data for most of the time. Taking advantage of this fact can lead to substantial speed-ups.

As an example of this consider an array in which one item is retrieved for 90% of the time. Instead of storing this item in the array an alternative is to store it in a separate variable. The operation which checks whether an item is in the array would examine this variable first, before executing what might be a lengthy search of the array. At the cost of an extra variable and a little code complication, a fast retrieval time can be achieved.

Another example of this technique in action occurs in the search of an array organized for a linear search. If a small subset of the data to be stored in the array is retrieved very frequently, then a good strategy is, once an item has been retrieved, to place it at the head of the array and shuffle the remaining items down by one location.

This would ensure that, for most of the time, the frequently accessed items will be at the head of the array and could be retrieved quickly. This strategy is quite a good one if you wish to implement a linear search with a few more items being stored than would normally be considered. At the expense of programming complexity, and some extra processing time during insertion, a simple linear search can be speeded up considerably.

13.3.3 Providing extra information about data location

Another popular technique, which sacrifices both space and programming complexity in order to achieve a high retrieval response time, is to embed extra data inside some stored data in order to provide clues and cues about where data to be accessed is stored.

The best example of this approach is the implementation of a set of employee records as an indexed array or indexed file. Assume that each record contains a large amount of employee details including items such as the name, address, tax code, and salary of the employee, together with a unique identification number. These records would be stored in a file or an array in some order. The order which would normally be chosen is that of ascending employee identification number.

Another array is also used to index this first array. Each entry in this array would contain an identification number of the employee, together with the location of the employee in the file. Normally only a small proportion

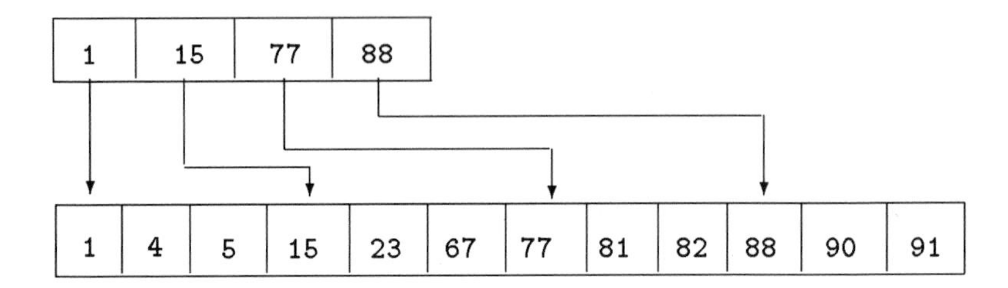

Figure 13.5 The use of an index

of the items in the first array are so indexed; in our example I shall assume that it is one in three. The organization of the data is shown in Figure 13.5. This means that when a particular item is to be retrieved the index is first searched, often using a binary search. The key nearest to the key searched for would be found, and this would give the position of the corresponding item in the array. A linear search can then be used to locate the item to be found.

Often the index can be stored in some fast access medium such as cache storage or main memory. This means that a small number of accesses are required to the medium on which the main data is stored—usually some very slow medium such as backing storage. If the data was stored without the index, but still in order, a binary search could have been used; however, it might have resulted in a large number of slow reads from a file device.

By organizing the data in such a way, with an index providing a clue to where the stored data is held, a faster response time is achieved. This is achieved at the expense of more complicated programming (insertion and deletion are more complex to program); extra memory to hold the index; and an increase in time for insertion and deletion operations.

A second example of the use of extra information is the implementation of a set as an ordered binary tree. The implementation of a set of integers is shown in Figure 13.6. An ordered binary tree is a structure which contains data and pointers to other data in the tree. The tree is ordered in the sense that if you examine a node in a tree you will find three fields. One field will contain the data stored, another field will contain a pointer to another tree which contains items which are less than the item at the node, and another field will contain a pointer to another tree which contains items greater than the item at the node.

The important property of such a storage structure is that access to the tree is very rapid, normally being proportional to $\ln n$, where n is the depth of the tree. Because the tree contains extra information about the location of other items of data, fast times can be achieved for insertion,

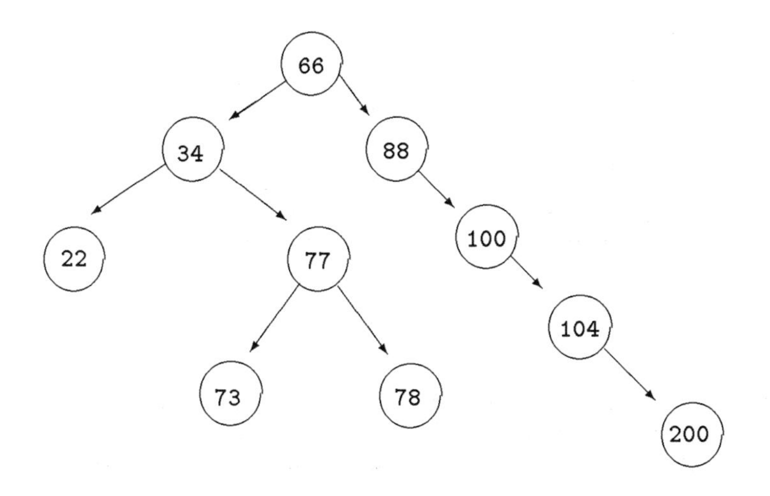

Figure 13.6 An ordered binary tree

deletion, and retrieval at the expense of quite a large overhead of memory and a slight complication in terms of programming.

The techniques that have been described so far in this section have all involved trading off memory and programming against speed of operations such as retrieval. Hardware is now so cheap that this is often the only trade-off a designer considers. However, before proceeding to the next section it it is worth giving one example of the reverse trade-off

13.3.4 Packing data

This technique was commonly used in the early days of computing where units of storage were in such short supply that they were crammed full of data. For example, you may have had an application in which the integers were no larger than 7 and where the computer had a word length of 24 bits. Such integers could easily be packed eight to a word at the expense of extra time for operations and some complexity in the programming.

13.4 The nature of design

In the early part of this book much stress has been made of the fact that the specifications that were presented did not contain any design or implementation directives. The specifications that have been presented have used data types such as sets, maps, and sequences which are not usually implemented in a programming language. Design, however, consists of implementing the operations of the specification in terms of operations on data types which are normally implemented in a programming language. Furthermore, the design will be based on many of the heuristics that were described in the previous section.

Normally a programming language will contain data types such as characters, integers up to a pre-defined limit, records, pointers, and arrays. The problem, then, is to design a system in terms of these data types, so that the software is a correct reflection of a specification that uses mathematical structures such as sets, sequences, and maps, at the same time employing heuristics which ensure that memory and time constraints are respected.

Many data types can be implemented easily. For example, characters in a top-level specification can be used directly in a design. A composite data type can be used directly and assumed to be implemented as a record. Integers on a computer can be modelled using natural numbers limited by a data invariant which specifies the upper and lower limit to the integer. Trees and graphs can be modelled in a design using composite data types and pointers, and sets can be modelled using arrays. This is the theme of the remainder of this chapter. In order to illustrate design principles it is worth looking at two examples of designs.

13.4.1 A set designed in terms of an array

The first design is that for the set specification shown below.

$[SYMBOLS]$

SymSys

$SymSet : \mathbb{P}\ SYMBOLS$

$\#SymSet \leq MaxSize$

ΔSymSys

$SymSys$
$SymSys'$

ΞSymSys

$\Delta SymSys$

$SymSet' = SymSet$

InitSet

$\Delta SymSys$

$SymSet' = \varnothing$

$$
\begin{array}{|l}
\multicolumn{1}{l}{\_\_ AddItem \_\_\_\_\_\_\_\_\_\_\_\_\_\_\_\_\_\_\_\_} \\
\Delta SymSys \\
sym? : SYMBOLS \\
\hline
\#SymSet < MaxSize \\
sym? \notin SymSet \\
SymSet' = SymSet \cup \{sym?\} \\
\hline
\end{array}
$$

$$
\begin{array}{|l}
\multicolumn{1}{l}{\_\_ RemItem \_\_\_\_\_\_\_\_\_\_\_\_\_\_\_\_\_\_\_\_} \\
\Delta SymSys \\
sym? : SYMBOLS \\
\hline
sym? \in SymSet \\
SymSet' = SymSet \setminus \{sym?\} \\
\hline
\end{array}
$$

$$
\begin{array}{|l}
\multicolumn{1}{l}{\_\_ CountItem \_\_\_\_\_\_\_\_\_\_\_\_\_\_\_\_\_\_\_\_} \\
\Xi SymSet \\
NumIn! : \mathbb{N} \\
\hline
NumIn! = \#SymSet \\
\hline
\end{array}
$$

This is a simple specification of a table of symbols which will contain no more than *MaxSize* symbols. Such a specification could, in a number of forms, be employed in system software or commercial data processing software, depending on what the symbols were. There are four operations specified: *InitSet* which initializes the set, *AddItem* which adds an item to the set, *RemItem* which removes an item, and *CountItem* which counts the number of items in the set.

One implementation of this set is of an array which is organized for a linear search. Such an array can be modelled by means of a sequence which has a data invariant that states that the maximum number of items in the sequence is some upper limit *MaxSize*. All an array is is a structure which associates a number i between its lower bound and upper bound with an item stored at its ith location.

$$\begin{array}{|l}
\hline
\textit{SymSysD} \\
\textit{SymSeq} : \text{seq}\,\textit{SYMBOLS} \\
\textit{NoSyms} : \mathbb{N} \\
\hline
\#\textit{SymSeq} = \textit{MaxSize} \\
\textit{NoSyms} \leq \textit{MaxSize} \\
\forall\, i, j : 1 \mathinner{..} \textit{NoSyms} \bullet \\
\quad i \neq j \Rightarrow \textit{SymSeq}\; i \neq \textit{SymSeq}\; j \\
\hline
\end{array}$$

$$\begin{array}{|l}
\hline
\Delta \textit{SymSysD} \\
\textit{SymSysD} \\
\textit{SymSysD}' \\
\hline
\end{array}$$

$$\begin{array}{|l}
\hline
\Xi \textit{SymSysD} \\
\Delta \textit{SymSysD} \\
\hline
\textit{SymSeq}' = \textit{SymSeq} \\
\textit{NoSyms}' = \textit{NoSyms} \\
\hline
\end{array}$$

It is worth looking at this specification in some detail. It models the set by means of a sequence of length *MaxSize* symbols. This means that every item in the sequence will be associated with a natural number ranging from 1 to *MaxSize*. This is exactly the property of a single-dimensional array. Also, the design includes a natural number *NoSyms* which will contain the number of elements in the array. This is an example of one of the design heuristics that were mentioned in the previous section: using an extra data item which saves continual re-calculation. This data item will contain the current number of symbols in the array. The system has now been annotated with the letter D to shown that it is a design.

The specification of the operations on this data which reflect the operations shown in the top-level specification are shown below. They are annotated with a capital letter D to show that they are now design operations.

$$\begin{array}{|l}
\hline
\textit{InitSetD} \\
\Delta \textit{SymSysD} \\
\hline
\textit{NoSyms}' = 0 \\
\hline
\end{array}$$

The *InitSetD* operation just sets the number of items in the array to zero. The specification for the *AddItemD* operation is shown below.

$$
\begin{array}{|l}
\hline
\textit{AddItemD} \\
\Delta \textit{SymSysD} \\
\textit{sym?} : \textit{SYMBOLS} \\
\hline
\textit{NoSyms} < \textit{MaxSize} \\
\textit{sym?} \notin \operatorname{ran} \textit{SymSeq} \lhd (1 \mathinner{\ldotp\ldotp} \textit{NoSyms}) \\
\textit{NoSyms}' = \textit{NoSyms} + 1 \\
\textit{SymSeq}\ \textit{NoSyms}' = \textit{sym?} \\
\textit{SymSeq}' \lhd (1 \mathinner{\ldotp\ldotp} \textit{NoSyms}) = \textit{SymSeq} \lhd (1 \mathinner{\ldotp\ldotp} \textit{NoSyms}) \\
\hline
\end{array}
$$

The first two predicates are the pre-condition: that there must be room for the symbol to be inserted in the array and that the symbol must not already be in the array. The final three predicates show that the number of items in the array has increased by one, the inserted symbol is placed at the end of the array, and the items in the array up to the inserted item remain the same.

This specification contains design details which constrain the implementor to one solution: that of adding the symbol to the array at the end of the array. A useful point can be made if an alternative version of this operation specification is presented. This is shown below.

$$
\begin{array}{|l}
\hline
\textit{AddItemD} \\
\Delta \textit{SymSysD} \\
\textit{sym?} : \textit{SYMBOLS} \\
\hline
\textit{NoSyms} < \textit{MaxSize} \\
\textit{sym?} \notin \operatorname{ran} \textit{SymSeq} \lhd (1 \mathinner{\ldotp\ldotp} \textit{NoSyms}) \\
\textit{NoSyms}' = \textit{NoSyms} + 1 \\
\operatorname{ran} \textit{SymSeq}' \lhd (1 \mathinner{\ldotp\ldotp} \textit{NoSyms}') = \\
\qquad \operatorname{ran} \textit{SymSeq} \lhd (1 \mathinner{\ldotp\ldotp} \textit{NoSyms}) \cup \{\textit{sym?}\} \\
\hline
\end{array}
$$

Here the bulk of this operation specification is the same as the previous version. However, one major component differs. The post-condition, instead of specifying that the $NoSym'$ item of the array is to be written to, just specifies that you will find the symbol that has been inserted in any one of the first $NoSym'$ positions in the array.

This leaves the implementer free to adopt more than one insertion strategy. It covers not only the placing of the symbol at the end of the array but also at the beginning or any other position in the array. Whether to leave design decisions open or not depends on a number of factors.

- The company who develops the software in which the symbol table is embedded may have a policy that their programmers (implementors) will receive prescriptive designs which have only one design option. For example, they may only employ junior or inexperienced staff as programmers who are incapable of carrying out design tasks.
- At this stage in the development of the software, information about the behaviour of the symbol table may not be at hand. For example, the designer may not yet know whether a few symbols are going to be retrieved. In this case the design decision is delayed.
- The design process may be carried out in a number of steps. That is, the design specifications of the operations may exist in a number of versions, each version becoming more prescriptive.

For the rest of this specification of the designed operations I shall assume that items will be added to the end of the array.

The specification of the operation *RemItemD* is shown below.

RemItemD

$\Delta SymSysD$
$sym? : SYMBOLS$

$sym? \in \operatorname{ran} SymSeq$

$NoSyms' = NoSyms - 1$

$\exists i : 1 \mathrel{..} NoSyms' \bullet$
$\quad SymSeq' \lhd (1 \mathrel{..} i - 1) \frown sym? \frown SymSeq' \lhd (i \mathrel{..} NoSyms)$
$\quad = SymSeq$

The first predicate is the pre-condition. It states that the symbol to be removed must be in the array. The final two predicates are the post-condition. The first describes the fact that the number of symbols in the array has decreased. The second describes the fact that the array no longer contains the symbol but that the ordering of the remaining elements in the array is the same. The final specification is that of *CountItemD*.

CountItemD

$\Xi SymSysD$
$count! : \mathbb{N}$

$count! = NoSyms$

All this specification states is that the result of the operation is that *count!* is set equal to the component of the data type which holds the current number of items in the array.

This, then, is a design for a top-level specification. It expresses, in mathematical terms, what should happen to a data type: the array, which reflects the facilities available in a programming language. The major question which we next need to address is: how do we check that the design is a correct implementation of the top-level specification?

13.5 Validating designs

One of the major research problems that remain for the formal methods community is to demonstrate that a design meets a high-level functional specification. The aim of this section is to briefly describe one promising avenue. The example that I shall use is the design of a structure known as a **bag**. A bag, or **multiset**, is a collection of values where repetition is allowed, as compared with a set where each element of a set occurs only once.

Z contains facilities for defining bags; [Spivey, 1989] contains a description. However, for the purposes of this chapter I shall assume that no such facility exists and that we need to define it.

The top-level specification of a bag of integers which are no larger than 100 is shown below.

```
── Bag ─────────────────────────
│ fbag : ℕ ⇸ ℕ₁
├──────────
│ ∀ i : dom fbag • i ≤ 100
└────────────────────────────────
```

The bag is implemented by means of a partial function which maps the bag elements into the number of times that they occur in the bag. Since the range of the function is the natural numbers exclusive of 1 then an integer which does not occur in a bag is not represented by an element in the function.

The schemas which define the data invariant and the fact that a bag remains unchanged are shown below.

```
── ΔBag ────────────────────────
│ Bag
│ Bag'
└────────────────────────────────
```

```
── ΞBag ────────────────────────
│ ΔBag
├──────────
│ fbag' = fbag
└────────────────────────────────
```

The following schemas define four operations on bags: an operation to initialize a bag, an operation to count the number of occurrences of an element in a bag, an operation to add an item to a bag, and an operation to delete an item from a bag.

$$\begin{array}{|l}
\multicolumn{1}{l}{\textit{InitBag}} \\
\Delta Bag \\
\hline
fbag' = \varnothing \\
\hline
\end{array}$$

$$\begin{array}{|l}
\multicolumn{1}{l}{\textit{NumBag}} \\
\Xi bag \\
elem? : \mathbb{N} \\
number! : \mathbb{N} \\
\hline
elem? \leq 100 \\
elem? \in \operatorname{dom} fbag \land number! = fbag\ elem? \\
\quad \lor \\
elem? \notin \operatorname{dom} fbag \land number! = 0 \\
\hline
\end{array}$$

$$\begin{array}{|l}
\multicolumn{1}{l}{\textit{AddBag}} \\
\Delta Bag \\
elem? : \mathbb{N} \\
\hline
elem? \leq 100 \\
elem? \notin \operatorname{dom} fbag \land fbag'\ elem? = 1 \\
\quad \lor \\
elem? \in \operatorname{dom} fbag \land fbag'\ elem? = (fbag\ elem?) + 1 \\
\hline
\end{array}$$

$$\begin{array}{|l}
\multicolumn{1}{l}{\textit{DelBag}} \\
\Delta Bag \\
elem? : \mathbb{N} \\
\hline
elem? \leq 100 \land fbag\ elem? \neq 0 \\
fbag\ elem? > 1 \land fbag'\ elem? = fbag\ elem? - 1 \\
\quad \lor \\
fbag\ elem? = 1 \land fbag' = fbag \ntriangleleft \{elem?\} \\
\hline
\end{array}$$

The schema *AddBag* has a pre-condition that the element to be added must be no greater than 100. The post-condition is split into two parts: the first handles the case that the element is not in the bag, the second handles the case that the element is in the bag.

The schema *DelBag* has a pre-condition which states that the element to be deleted must be no bigger than 100 and that the element is actually in the bag. The post-condition has two components: the first component is true when the element to be deleted occurs more than once in the bag; the second component is true when the element to be deleted occurs just once.

This, then, is the top-level specification of a bag. The next stage is to derive a data type nearer to the data structures in a programming language. The data type that I will choose is the obvious one: that of an array which has bounds from 0 to 100 and whose ith element will contain the number of occurrences of i in the bag. This is quite a good choice if the bag is not sparse, that is, it will contain a large number of individual elements. The schema describing the array is shown below.

BagD
$$bagarr : \mathbb{N} \twoheadrightarrow \mathbb{N}$$

$$\forall i : \mathrm{dom}\, bagarr \bullet i \leq 100$$
$$\# bagarr = 101$$

The first predicate is the same as that for the top-level specification. The second predicate states that the function will always have a domain which will contain 101 elements and since the domain starts at 0 and finishes at 100, this implies that the domain will be *all* the natural numbers between these bounds. The schemas which define the invariant and the fact that the data type is unchanged are shown below:

$\Delta BagD$
$$BagD$$
$$BagD'$$

$\Xi BagD$
$$\Delta BagD$$

$$bagarr' = bagarr$$

The link between a top-level specification and a design specification is established by means of a predicate known as a **coupling invariant**. This is a predicate which relates the top-level state and the design state and which remains true throughout the execution of a system. In the bag example the coupling invariant is

$$\begin{aligned}&\forall\, i : \operatorname{dom} bagarr \bullet \\ &\quad bagarr\ i > 0 \land i \in \operatorname{dom} fbag \land bagarr\ i = fbag\ i \\ &\quad \lor\ bagarr\ i = 0 \land i \notin \operatorname{dom} fbag\end{aligned}$$

All this states is that if you examine the ith element of the array $bagarr$ then, if it is zero, i will not occur in the domain of $fbag$. Also, if the element is greater than zero then it will occur in the domain of $fbag$ and the ith element of $bagarr$ will be $fbag\ i$. The coupling invariant can be used to derive the pre and post-conditions of the designed operations which correspond to the top-level operations *InitBag*, *NumBag*, *DelBag*, and *AddBag*. This is carried out using the coupling invariant to eliminate references to the top-level data type. In order to demonstrate this consider the schema which describes the initialization of a bag.

```
┌─ InitBag ──────────────────────────
│ ΔBag
├──────────
│ fbag' = ∅
└────────────────────────────────────
```

Since the coupling invariant will remain true throughout the exection of the program code which implements bags we can say that

$$\begin{aligned}&\forall\, i : \operatorname{dom} bagarr' \bullet \\ &\quad bagarr'\ i > 0 \land i \in \operatorname{dom} fbag' \land bagarr'\ i = fbag'\ i\ \lor \\ &\quad bagarr'\ i = 0 \land i \notin \operatorname{dom} fbag'\end{aligned}$$

Thus, all the following predicates are true:

1 $\quad fbag' = \varnothing$

2 $\quad \forall\, i : \operatorname{dom} bagarr' \bullet$
$\qquad bagarr'\ i > 0 \land i \in \operatorname{dom} fbag' \land bagarr'\ i = fbag'\ i$
$\qquad \lor\ bagarr'\ i = 0 \land i \notin \operatorname{dom} fbag'$

The first disjunct in the quantified expression will always be false since predicate 1 states that $fbag'$ is the empty set. This means that the predicate

3 $\quad \forall\, i : \operatorname{dom} bagarr' \bullet$
$\qquad bagarr'\ i = 0 \land i \notin \operatorname{dom} fbag'$

is true. Now since from predicate 1 $fbag'$ is the empty set, the second conjunct in the quantified expression in 3 is always true. This gives the predicate

$$\forall\, i : \operatorname{dom} bagarr' \bullet bagarr'\ i = 0$$

thus transforming the schema

$$\begin{array}{|l}\multicolumn{1}{l}{\underline{\ InitBag\ \qquad\qquad\qquad\qquad\qquad\qquad\qquad\qquad\qquad}}\\ \Delta Bag \\ \hline fbag' = \varnothing \\ \hline\end{array}$$

to

$$\begin{array}{|l}\multicolumn{1}{l}{\underline{\ InitBagD\ \qquad\qquad\qquad\qquad\qquad\qquad\qquad\qquad}}\\ \Delta BagD \\ \hline \forall\, i : \operatorname{dom} bagarr' \bullet bagarr'\ i = 0 \\ \hline\end{array}$$

Thus, by means of the coupling invariant we have derived the schema for the operation *InitBagD*. The derivation of the remaining designed operations is similar.

The derivation of *NumBagD* follows. First, we have the schema for *NumBag*.

$$\begin{array}{|l}\multicolumn{1}{l}{\underline{\ NumBag\ \qquad\qquad\qquad\qquad\qquad\qquad\qquad\qquad\qquad}}\\ \Xi Bag \\ elem? : \mathbb{N} \\ number! : \mathbb{N} \\ \hline elem? \leq 100 \\ elem? \in \operatorname{dom} fbag \wedge number! = fbag\ elem? \\ \qquad \vee \\ elem? \notin \operatorname{dom} fbag \wedge number! = 0 \\ \hline\end{array}$$

This means that the following predicates are true:

$$\begin{array}{ll} 1 & fbag = fbag' \\ 2 & elem? \leq 100 \\ 3 & \forall\, i : \operatorname{dom} bagarr' \bullet \\ & \qquad bagarr'\ i > 0 \wedge i \in \operatorname{dom} fbag' \wedge bagarr'\ i = fbag'\ i \\ & \qquad\qquad\qquad \vee \\ & \qquad bagarr'\ i = 0 \wedge i \notin \operatorname{dom} fbag' \end{array}$$

with the first predicate being taken from the schema ΞBag.

The predicate which needs to be transformed is the post-condition of *NumBag*:

$$\begin{array}{l} elem? \in \operatorname{dom} fbag \wedge number! = fbag\ elem? \\ \qquad \vee \\ elem? \notin \operatorname{dom} fbag \wedge number! = 0 \end{array}$$

This can first be transformed to predicate 4:

$$4 \quad elem? \in \mathrm{dom}\, fbag' \wedge number! = fbag'\ elem? \\ \vee \\ elem? \notin \mathrm{dom}\, fbag' \wedge number! = 0$$

because of the truth of predicate 1. Since *elem*? is in the domain of *bagarr*′, as it is less than or equal to 100, we can use 3 to write predicate 5 below

$$5 \quad bagarr'\ elem? > 0 \wedge elem? \in \mathrm{dom}\, fbag' \wedge bagarr'\ elem? = fbag'\ elem? \\ \vee \\ bagarr'\ elem? = 0 \wedge elem? \notin \mathrm{dom}\, fbag'$$

The next step is to carry out a case analysis on the predicates 4 and 5. First, assume that $elem? \in \mathrm{dom}\, fbag'$. This gives predicate 6:

$$6 \quad number! = fbag'\ elem? \wedge bagarr'\ elem? > 0 \wedge \\ bagarr'\ elem? = fbag'\ elem?$$

from which we can extract predicate 7:

$$7 \quad number! = bagarr'\ elem?$$

Next, we assume that $elem? \notin \mathrm{dom}\, fbag'$. Predicates 4 and 5 give predicate 8

$$8 \quad number! = 0 \wedge bagarr'\ elem? = 0$$

From this predicate can be extracted predicate 9

$$9 \quad number! = bagarr'\ elem?$$

Predicate 7 covers the case that $elem? \in \mathrm{dom}\, fbag'$ and predicate 9 covers the case that $elem? \notin \mathrm{dom}\, fbag'$. Since predicates 7 and 9 are the same we can assert that the predicate $number! = bagarr'\ elem?$ will always be true and hence the schema for *NumBagD* can be written as

```
┌─ NumBagD ──────────────────────────
│ ΞBagD
│ elem? : ℕ
│ number! : ℕ
├────────────
│ elem? ≤ 100
│
│ number! = bagarr' elem?
└────────────────────────────────────
```

The next operation to refine is *AddBag*. Its definition is shown below:

$$
\begin{array}{|l}
\multicolumn{1}{l}{\underline{\quad AddBag \quad}} \\
\Delta Bag \\
elem? : \mathbb{N} \\
\hline
elem? \leq 100 \\
elem? \notin \operatorname{dom} fbag \land fbag'\ elem? = 1 \\
\quad \lor \\
elem? \in \operatorname{dom} fbag \land fbag'\ elem? = (fbag\ elem?) + 1 \\
\hline
\end{array}
$$

This, together with the coupling invariant, gives the following predicates:

1 $\quad elem? \leq 100$

2 $\quad \forall i : \operatorname{dom} bagarr \bullet$

$$
\begin{array}{l}
bagarr\ i > 0 \land i \in \operatorname{dom} fbag \land bagarr\ i = fbag\ i \\
\quad \lor \\
bagarr\ i = 0 \land i \notin \operatorname{dom} fbag
\end{array}
$$

3 $\quad elem? \notin \operatorname{dom} fbag \land fbag'\ elem? = 1$

$$
\begin{array}{l}
\quad \lor \\
elem? \in \operatorname{dom} fbag \land fbag'\ elem? = (fbag\ elem?) + 1
\end{array}
$$

The aim as with the previous refinements is to find the post-condition of an operation *AddBagD*. Since predicate 1 tells us that *elem?* is in the domain of *bagarr* and *bagarr′* predicate 2 can be transformed to predicate 4.

4 $\quad bagarr\ elem? > 0 \land elem? \in \operatorname{dom} fbag \land bagarr\ elem? = fbag\ elem?$

$$
\begin{array}{l}
\quad \lor \\
bagarr\ elem? = 0 \land elem? \notin \operatorname{dom} fbag
\end{array}
$$

In order to obtain a predicate involving *bagarr* and *elem?* a case analysis is necessary. First, assume that $elem? \in \operatorname{dom} fbag$. This means that predicate 4 can be written as predicate 5:

5 $\quad bagarr\ elem? > 0 \land bagarr\ elem? = fbag\ elem?$

and predicate 3 can be written as predicate 6.

6 $\quad fbag'\ elem? = (fbag\ elem?) + 1$

The final conjunct of predicate 5 and predicate 6 gives predicate 7:

7 $\quad fbag'\ elem? = (bagarr\ elem?) + 1$

The next case analysis is for $elem? \notin \operatorname{dom} fbag$. Predicate 4 gives predicate 8:

8 $\quad bagarr\ elem? = 0$

Predicate 3 can be written as predicate 9:

9 $fbag'\ elem? = 1$

By writing predicate 9 as $fbag'\ elem? = 0 + 1$ predicates 8 and 9 can be combined to give the predicate 10:

10 $fbag'\ elem? = bagarr\ elem? + 1$

which is the same as the predicate derived from the case $elem? \in \operatorname{dom} fbag$, hence the above predicate is true. All that remains is to remove the reference to $fbag'$ from this predicate. Since the coupling invariant will always be true it can be primed, and since $elem? \in \operatorname{dom} bagarr'$ we can write it as predicate 11:

11 $bagarr'\ elem? > 0 \land elem? \in \operatorname{dom} fbag' \land bagarr'\ elem? = fbag'\ elem?$
$\lor$
$bagarr'\ elem? = 0 \land elem? \notin \operatorname{dom} fbag'$

Predicate 10 implies that $elem? \in \operatorname{dom} fbag'$ and hence predicate 11 can be reduced to

$$bagarr'\ elem? > 0 \land bagarr'\ elem? = fbag'\ elem?$$

Using the second conjunct of this predicate to simplify predicate 10 gives

$$bagarr'\ elem? = bagarr\ elem? + 1$$

Hence the schema for *AddBagD* can be written as

```
┌─ AddBagD ──────────────────────────
│ ΔBagD
│ elem? : ℕ
├──────────
│ elem? ≤ 100
│ bagarr′ elem? = bagarr elem? + 1
└────────────────────────────────────
```

The schema for the *DelBagD* operation can be derived in a similar way.

13.6 Summary

This chapter has been one of the key ones in the book. It has examined the nature of design and described a number of common design heuristics. It has shown that a system can be designed by employing techniques such as adding redundant data. An important point that has been described in this chapter is that the same language used for top-level specifications can be used to describe stored data and operations in a design. The chapter continued with a description of two examples of top-level specifications and their designs. Finally, a technique for demonstrating equivalence between a top-level specification and its design was presented.

13.7 Further reading

One of the best books about software design which masquerades under a title which suggests it is about hacking is [Bentley, 1982]. It contains much advice on design tactics and source code level optimizations. It contains a number of case studies and is an excellent example of what a well-written book should be. Many of the design heuristics described in this book were taken from it.

Bibliography

[Andrews and Ince, 1991] Andrews, D and Ince, D (1991), *Practical Formal Methods with VDM*, McGraw-Hill, Maidenhead.

[Bell *et al.*, 1977] Bell, T E, Bixler, D C, and Dyer, M (1977), 'An extendable approach to computer aided software requirements engineering', *IEEE Transactions on Software Engineering*, **3**(1): 49–60.

[Bentley, 1982] Bentley, J L (1982), *Writing Efficient Programs*, Prentice-Hall, Englewwod Cliffs, NJ.

[Bjorner and Jones, 1982] Bjorner, D and Jones, C B (1982), *Formal Specification and Software Development*, Prentice-Hall, Englewood Cliffs, NJ.

[Boehm, 1981] Boehm, B W (1981), *Software Engineering Economics*, Prentice-Hall, Englewood Cliffs, NJ.

[Cain and Gordon, 1975] Cain, S H and Gordon, E K (1975), 'PDL–a tool for software design', in *Proceedings National Computer Conference*, **44**, 211–76.

[Clarke, 1976] Clarke, L (1976), 'A system to generate test data', *IEEE Transactions on Software Engineering*, **2**(3): 215–222.

[Cohen *et al.*, 1986] Cohen, B, Harwood, W T, and Jackson, M I (1986), *The Specification of Complex Systems*, Addison-Wesley.

[Constable and O'Donell, 1978] Constable, R L and O'Donell, M J (1978), *A Programming Logic*, Winthrop, Cambridge, Mass.

[Craigen and Summerskill, 1990] Craigen, D and Summerskill, K (eds) (1990), *Formal Methods for Trustworthy Computer Systems(FM89)*, Springer-Verlag, New York, NY.

[Davis, 1990] Davis, A M (1990), *Software Requirements, Analysis and Specification*, Prentice-Hall, Englewood Cliffs, NJ.

[Diller, 1990] Diller, A (1990), *Z An Introduction to Formal Methods*, Wiley, Chichester.

[Empson, 1977] Empson, W (1977), *Seven Types of Ambiguity*, Penguin Books, London.

[Gerhart *et al.*, 1980] Gerhart, D L, Muser, D R, *et al.* (1980), 'An overview of AFFIRM. A specification and verification system', in *Proceedings IFIP Congress*, North-Holland, Amsterdam.

[Guttag and Horning, 1978] Guttag, J V and Horning, J J (1978), 'The algebraic specification of abstract data types', *Acta Informatica*, **10**(1): 27–52.

[Humphrey, 1989] Humphrey, W S (1989), *Managing the Software Process*, Addison-Wesley, Wokingham.

[Ince, 1988] Ince, D C (1988), *Software Development: Fashioning the Baroque*, Oxford University Press, Oxford.

[Kolman and Busby, 1987] Kolman, B and Busby, R C (1987), *Discrete Mathematical Structures*, Prentice-Hall, Englewood Cliffs, NJ.

[Luckham *et al.*, 1979] Luckham, D, German, S, *et al.* (1979), *Stanford Pascal Verifier user's manual*, Technical report CS-79-731, Computer Science Department, Stanford University.

[Meyer, 1985] Meyer, B (1985), 'On formalism in specifications', *IEEE Computer*, **18**(1): 6–26.

[Miller, 1977] Miller, E F (1977), 'Program testing: art meets theory', *IEEE Computer*, **10**(6): 42–51.

[Mills, 1988] Mills, H D (1988), *Software Productivity*, Dorset, New York, NY.

[Osterweil, 1983] Osterweil, L J (1983), 'Toolpack—an experimental software development environment research project', *IEEE Transactions on Software Engineering*, **9**(6): 673–686.

[Potter *et al.*, 1991] Potter, B, Sinclair, J, and Till, D (1991), *An Introduction to Formal Specification and Z*, Prentice-Hall, Hemel Hempstead.

[Sacks, 1985] Sacks, O (1985), *The Man who Thought his Wife was his Hat*, Picador, London.

[Schoman and Ross, 1977] Schoman, K and Ross, D T (1977), 'Structured analysis for requirements definition', *IEEE Transactions on Software Engineering*, **3**(1): 6–15.

[Spivey, 1989] Spivey, J M (1989), *The Z Notation*, Prentice-Hall, Hemel Hempstead.

[Stoll, 1961] Stoll, R R (1961), *Sets, Logic and Axiomatic Theories*, Freeman, New York, NY.

[Teichrow and Hershey, 1977] Teichrow, D and Hershey, E A (1977), 'PSL/PSA: a computer aided technique for structured documentation and analysis of information processing systems', *IEEE Transactions on Software Engineering*, **3**(1): 41–48.

[Weinberg, 1980] Weinberg, V (1980), *Structured Analysis*, Prentice-Hall, Englewood Cliffs, NJ.

[Yourdon, 1989] Yourdon, E (1989), *Modern Structured Analysis*, McGraw-Hill, New York, NY.

Formal Specification using Z. D. Lightfoot (1991) MacMillan.

Software Development with Z:
A Practical Approach to Formal Methods
in Software Engineering. J.B. Wordsworth (1992)
Addison-Wesley.

IEEE Software - Sept. 1990 Issue

Computer Journal - Vol. 35, No. 5 (1992)
Formal Methods: Use & relevance for the
Development of Safety Critical Systems
L.M. Barocca & J.A. McDermid

Index